S0-APM-975

Fields
of
Fire

**An Illustrated History
of
Canadian Petroleum**

with over 100 photos

David Finch and Gordon Jaremko

Detselig Enterprises Ltd.
Calgary, Alberta

Fields of Fire: An Illustrated History of Canadian Petroleum

© 1994 David Finch and Gordon Jaremko

Canadian Cataloguing in Publication Data

Finch, David, 1956–
　Fields of fire

Includes index.
ISBN 1-55059-087-1

1. Petroleum industry and trade–Canada–History.　I.
Jaremko, Gordon.　II. Title.
HD9574.C22F56　1994　　338.2′7282′097109　　C94-910587-2

Detselig Enterprises Ltd. appreciates the financial support for
our 1994 publishing program, provided by the Department of
Communications, Canada Council and the Alberta Foundation
for the Arts, a beneficiary of the Lottery Fund of the Government
of Alberta.

Detselig Enterprises Ltd.
210, 1220 Kensington Road NW
Calgary, Alberta T2N 3P5

Edited by Sherry Wilson McEwen

Cover design by Dean MacDonald

Front cover photo courtesy of Husky Oil

Back cover photo courtesy of Petro-Canada

Printed and bound in Hong Kong

ISBN 1-55059-087-1　　　　　　　　　　SAN 115-0324

Contents

Oil stock promotions along Stephen Avenue, Calgary, May 1914
(Harry Pollard/Provincial Archives of Alberta/P1306)

Black Gold

Imagine doing without oil and natural gas. Until at least the mid-1980s, it was still possible to glimpse such a life without leaving Canada. In Alberta, the home of the petroleum industry, a taste of getting by without oil could be had at the last frontier of old-fashioned homesteading, a 450-kilometre belt along the Peace River.

After paying a call on the sodbusters who live there, you learned why places are given names like "Breaking Point" locally, even though they may not be noted this way on official maps. Coming across such a sign, posted on a rough back road, foretells the tale of early Northern homesteading, where plows broke virgin soil, and where the land sometimes broke the tools, bank accounts, back and hearts of the pioneers. And it continues to modern times.

"You break it or it breaks you," homesteader Gilbert Ducharme explained to visitors, pointing to an abandoned house filled with snow in a patch of land still half-covered in bush and debris. Fuel, and especially the scarcity and cost of it beyond the reach of service-stations chains and heating-utility networks, played a big role in the struggle. It cost up to $100 an hour, for example, to hire the mechanical muscle needed to create new crop fields by clearing away the trees and stumps.

The alternative to paying for a diesel-guzzling Caterpillar tractor was months or even years of hard labor with an axe. Most homesteaders combined the two methods, clearing by hand while saving up to hire Cats for the hardest parts. Ducharme's neighbors, Stan and Cathy Marsh, made their son and three daughters earn their allowances by digging up tree roots.

Household comforts often come last in a society that had to struggle for fuel needed for heavy lifting jobs. Even well-established frontiersmen like Lawrence, Frederick and Phillip Lambert knew the grain of truth in the old joke about firewood. It warms its users twice, the first time when they cut it.

Veterans of the homestead frontier settled for room temperatures that would leave most city dwellers freezing. They hardened to the point where it was routine to don a down-filled vest for a brisk session at a chopping block in -30°C weather. Heat is a high-value resource in such regions, for good reason. A city house equipped with a natural-gas furnace uses 3.4 cubic metres per hour and the hot-water heater uses 0.9 cubic metres per hour. A pioneer would have to cut firewood at a back-breaking pace of about 1.5 one-ton cords per month to match this energy consumption.

The homesteaders also proved that there is more to oil and natural gas than just convenience and labor-saving. These precious resources improve safety too. Like the Marshes, Lawrence Lambert had to rebuild a house that burned down due to problems with old-fashioned stoves, pipes and construction materials. He could not replace the wife, four sons and a daughter who died in the inferno that levelled his first house.

As a result of such tragedies, the homesteading frontier invented the "fire shower," where neighbors held parties to collect food, clothing, furniture and money for families whose homes burned down. The Alberta government officially recognized the risks of old-fashioned ways with a program called "Disaster Services." Disaster Services made trailer-style portable housing available quickly to fire victims who agreed to pay for the units on a rent-to-own, interest-free instalment plan.

The potent combination of power, comfort and safety adds up to make oil and its cousin, natural gas, more than "black gold" – the old term of a source of wealth for a few. A society that grows accustomed to oil and gas rapidly finds them to be essentials, required in bulk. Rather than representing fortunes for a few, they create livelihoods for many.

Just as Alberta's homestead frontier continues to epitomize what life can be without oil and gas, the province as a whole demonstrates how finding oil and gas in bulk pays off, even after eight years of lean times brought on by weak energy prices.

Although partially recovered, oil still fetches 20 percent less than the top price, $29.75 a barrel. This price was set by the National Energy Program in the months before its Liberal architects lost power and their Conservative heirs dismantled the NEP in 1984-85. Even in the worst depths of the drilling slumps, corporate staff-cutting and job-skill mergers, oil was responsible for 70 percent of all goods produced in the province and half the provincial income. (Source: A 1991 survey by the Calgary economics consulting firm of Wright Mansell Research Ltd., *The Role of the Oil and Gas Industry in the Alberta Economy.*)

Without oil, it is often said, Alberta would be Saskatchewan – sparsely populated and reliant on global grain markets that have become a byword for gluts and instability. The economists checked census records and found the claim makes sense. Until the 1947 Leduc discovery made Alberta an oil and gas producer on the international scale, Saskatchewan was a bigger province.

As late as 1946, Saskatchewan had 833 000 residents while Alberta had 803 000. Alberta started pulling ahead by 1951, when its population rose to 940 000 while Saskatchewan stagnated at 832 000. After four decades of this consistent trend, Alberta's population reached 2.5 million in 1991 while Saskatchewan had 995 000 residents. As of 1989, personal income in Alberta was $21 075 per capita compared to $17 109 in Saskatchewan.

Oil and gas, the economists concluded, amount to a "propulsive or motor industry." As it grows up producing commodities that come to be regarded as essentials, such an industry becomes an agent of change for an entire society. It not only creates new wealth, it has spin-offs on a large scale. These include requirements for high levels of investment, labor training, management skills and sophisticated suppliers of services such as transportation, construction, engineering and computers.

Until the 1986 global oil-price crash and the slower but equally drastic collapse of North American natural-gas prices forced corporate and government thinking about the fundamentals of energy economics to change, the Canadian petroleum industry had no trouble earning its reputation as an engine of growth. From humble beginnings in Turner Valley, it grew to dominate Alberta after Leduc. By the 1960s, the industry outgrew conventional drilling to start building mining and processing complexes the size of cities in the oil sands.

At the same time, the industry gradually expanded far beyond its Alberta base into the Northwest Territories and out to sea offshore of Atlantic Canada. Even the NEP, reviled in Alberta when it was introduced in 1980, did not stall the engine. It only changed directions, with a combination of tax, grant and pricing policies that lit fires under northern and offshore activity at the expense of the industry's home base.

In hindsight, it has become all too easy to dismiss the impulse that drove the often hugely expensive attempts to expand as economic naivete or wild-eyed optimism. At the time, the industry's long-term fundamentals told oil and gas strategists that the world wanted as much as they could find and would cover the costs. The best economic minds, outside as well as inside the petroleum industry, found it difficult to conceive of oil losing its commanding role. Since Winston Churchill started converting Britain's Royal Navy to oil from coal around the turn of the century, oil had gradually won a place as strategic, as well as useful, supplier of commodities.

In the mid-1970s, the Organization of Petroleum Exporting Countries (OPEC) continually raised prices to the point where demand for oil faltered, then fell, while simultaneously stimulating new production from outside the cartel. At the time, influential international economists like Dankwart Rustow and John Mungo – in "OPEC: Success and Prospects" (*Council on Foreign Relations*. New York: New York University Press, 1976) – considered the possibility that OPEC might be sowing the seeds of its own decay and rejected it. They concluded "OPEC seems to us . . . stronger and more durable than many competent observers have allowed."

Back then, long-term trends that drove market projections, the bread and butter of economists, told them that industrialization was spreading; that it relied on oil; and that oil was scarce. In the decades that followed World War II, the economies outside the Communist blocs recorded the greatest growth ever. Trade expanded five-fold. The value of goods and services produced tripled. Energy consumption likewise tripled.

From a tiny start at 40 000 in 1950, oil imports by the leading industrial countries multiplied. Japan's purchases proliferated by 135-fold, to 5.4 million barrels per day by the mid-1970s. Western European imports rose as high as 15.4 million barrels a day from 1.2 million. As U.S. oil fields, which were as much as a century old, depleted, Canada's neighbor and biggest trading partner also became a major oil importer. Between 1950 and the mid-1970s, U.S. imports nearly doubled to exceed 8 and sometimes 9 million barrels a day.

As an oil exporter and owner of comparatively unexplored frontiers, it came naturally to Canada to try cashing in on global trends. Anyone who lived in Alberta in the 1970s found the following kind of talk familiar and congenial: "Hydrocarbon resources are limited and exhaustible, and their proper exploitation determines . . . economic development. Foreign capital . . . can play an important role." The statements came from an OPEC policy statement, but could just as easily have come from Peter Lougheed as Alberta's premier – or the federal energy ministers who devised incentives for Arctic and offshore exploration.

The 1985 Western Accord on Energy and then the price collapse of 1986 put a stop to high rolling on the industry's technological and geographical frontiers. But just as low-profile, modest projects have steadily increased oil sands output, the North and Atlantic Canada remain far from forgotten. In June of 1994, following reappraisals of resource inventories and geological prospects, the federal government issued the first invitation in 25 years for bids on drilling rights in the southwestern part of the Northwest Territories.

On the East Coast, Petro-Canada is far from alone in suggesting that the oil industry will carry out the political promise behind government aid in constructing the Hibernia production platform - as the foundation for a "new petroleum province."

Mobil Oil Canada Ltd. plainly thinks the same way. As a wholly-owned subsidiary of Mobil Corp., which is based in Fairfax, VA., the company has no Canadian shareholders to demand reports and no need to tout its projects. But in internal annual reports to its employees, Mobil Oil Canada has made it clear that it is far from finished with

Atlantic Canada. The staff has been told a 10-year, "integrated strategy" crafted in 1990 looks out to sea.

"Our new strategy focuses on the east coast, an area where Mobil has been an industry leader since the mid-1960s. This area encompasses our existing discoveries at Hibernia, offshore Newfoundland and Sable Island offshore of Nova Scotia."

Mobil, like its peers, adds that "in today's economic climate, we cannot afford to explore with the simplistic objective of finding oil and gas. Our exploration activities have to be both economic and market driven."

And the post price-collapse era calls for "a broader business insight into exploration." Offshore, that means finding "giants," oil and gas pools so big that the unit costs of development and production stay low. Even in 1990, a year of apparently complete inactivity in Atlantic Canada, Mobil quietly acquired 560 000 hectares of prospects on the Grand Banks near Hibernia and on the west side of Newfoundland, which expanded its inventory of drilling prospects to 14 from 9. The Canadian industry likes to call itself "mature" now, and accordingly will move more slowly – but it is also obviously far from finished.

Our Thanks to . . .

To write history or journalism is to learn there is no such thing as the last word. Each book or article generates new questions. The question that propelled this book – what did things *feel* and *look* like during the oil industry's developing years – came soon after the authors participated in writing a narrative overview for the Petroleum History Society (McKenzie-Brown, Jaremko and Finch. *The Great Oil Age*. Calgary: Detselig, 1994).

It is history's good fortune that the oil industry has always had a sense that development has a wider, more historic significance than simply making money. Leading companies had pictures taken of their work, often by first-class professional photographers, and have preserved them through the boom-'n-bust cycles of energy economics. This volume – and the work of future historians – owes much to the executives and public affairs personnel whose responsibilities include custody of their pictorial treasure. They let us paw through these photos and allowed us to borrow some of the best images.

We thank: Noel Clelland at Sproule Associates Ltd., Gray Alexander at Panarctic Oils Ltd., Pius Rolheiser of Imperial Oil Ltd., Judy Wish at Petro-Canada, Heather Douglas at Mobil Oil Canada Ltd., Brenda Watson of Gulf Canada Resources Ltd., Peter Marshall of Syncrude Canada Ltd., Dave Ryan at Suncor Inc., Dawn Mitchell (when she was with Husky Oil Ltd.) and Bruce Martin of Canadian Marine Drilling Ltd. The historic images come from even more varied sources, but most of them now reside in the photo collections at the Provincial Archives of Alberta in Edmonton and at the Glenbow-Alberta Institute Archives in Calgary.

The text of this volume largely relies on primary sources, and often on interviews conducted with participants in the industry of all ranks over more than a decade. Rather than clutter the pages with footnotes, references to secondary sources have been incorporated into the text and a section, "Further Reading," has been added to the back. For a more complete bibliography of the industry, see *The Great Oil Age*.

And finally, our thanks to our editor, Sherry Wilson McEwen. Her helpful comments, suggestions and other editing skills created a product we are all proud to call our own. Many thanks also to Ted Giles who encouraged and supported us as we worked on this illustrated review of the Canadian petroleum industry.

Gushers, Dry Holes and Blowouts

. . . early oil finds

Oil flowing from tar sands deposit, Alberta n.d.
(Provincial Archives of Alberta/A12182)

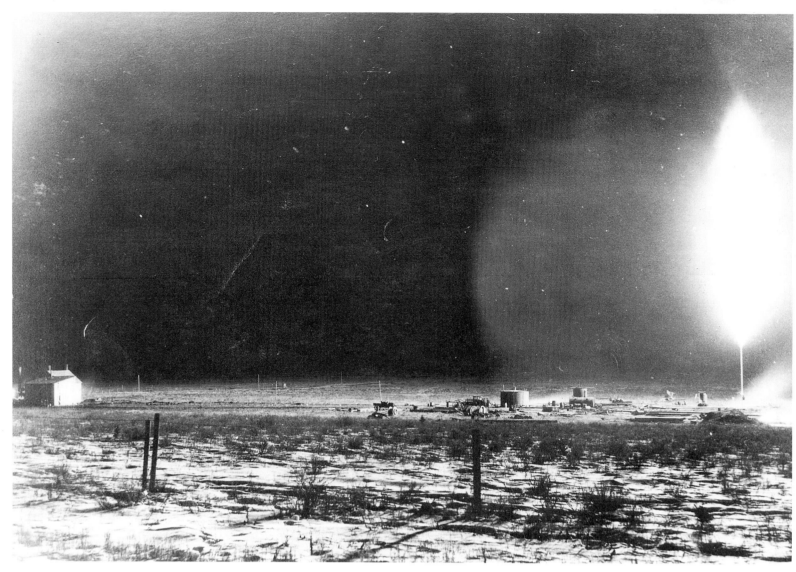

Royalite No. 4 well blowout, Turner Valley, November 1924
(Glenbow Archives/NA-701-9)

Royalite No. 4 well under control, Turner Valley, c. 1925 n.d.
(W.J. Oliver/Glenbow Archives/ND-8-429)

"I seen the ball of fire go from the stack on the boiler and go through the air and set the gas on fire."

That's how a nine-year-old named Don remembered the Royalite No. 4 well catching fire. The explosive events of that Sunday in October 1924 stuck with Don for the rest of his life. His father, Sam Coultis, was in charge of Royalite's operations in the Turner Valley oil field southwest of Calgary when the blowout torched the rig. Weeks later, wild well fighters from Oklahoma used steam and dynamite to kill the inferno.

There are discoveries, and then there are events that put a place on the map. Royalite No. 4 was the well that made the Western Canadian oil field of Turner Valley famous. This wild well, once extinguished and capped, provided enormous amounts of natural gas. It proved the field's value. Royalite built a scrubbing plant to remove the noxious and smelly sulphur, then sent the valuable gas to Calgary by pipeline. Turner Valley produces natural gas to this day.

Most discoveries are not nearly so spectacular or profitable. Many others were just teasers, tantalizing glimpses of the potential profits of petroleum. For centuries, native peoples, explorers, trappers and traders regularly reported seeing oil and gas seepages all across Canada.

From our late 20th century perspective, these early discoveries are fascinating. From the perspective of the early settlers and explorers, the gooey, gummy, smelly or explosive oil and gas was merely a puzzling nuisance. While discovering resources is simple, the genius comes in finding a way to market commonplace items like petroleum, natural gas, coal, wood or fish.

And turning a natural phenomenon into a natural resource is not always apparent or feasible. Only a few centuries ago, beaver pelts were so valuable that the Hudson's Bay Company used them as the basis for building an empire. Today, with virtually no market for their fur, these busy rodents attract our attention only when they flood a creek or their survival is endangered.

In the 1700s, the Hudson's Bay Company made it a practice to exploit natural resources but even it could not find a way to market petroleum. While working for the great trading empire, Cree native Wa-pa-su travelled from Fort York to the Athabasca area, pursuing furs. In 1719 he returned with a lump of oily sand from the Athabasca River. When he presented the sticky curiosity to Henry Kelsey at Fort York, the Hudson's Bay Company official was unimpressed. A fastidious record keeper, he kept a detailed diary. On June 12, 1719 he mentioned the receipt of a sample of "the Gum or Pitch that flows out of the Banks of that River."

There were few uses for the tarry substance in those early days. Another fur trader, Peter Pond, stumbled onto the pitch in 1778 while travelling down the Clearwater River to the Athabasca. A feisty American explorer of ill repute, Pond noted "springs of bitumen that flow along the ground."

In the 1780s, British explorer Alexander Mackenzie passed through the area and reported seeing Natives "pitching" their canoes with the gum that flowed from the river banks. And so the discovery of oil in North America was anticlimactic. No rewards, medals or knighthoods for these early oil scouts. Fur traders perhaps used the gum to waterproof their canoes, but other than the occasional mention in a journal, society ignored oil.

To this day, the tar sands of northern Alberta remain largely undervalued. The sands do not give up their wealth willingly. Attempts to extract petroleum from the Athabasca tar sands by injecting steam into the ground failed in 1910. During the mid-1940s, the Abasand company met limited success when it attempted to develop the first commercial tar sands plant in Canada. Desperation set in at one point in the late 1950s and developers considered detonating a nuclear device deep in the Fort McMurray sands.

Luckily, cooler heads prevailed and in 1967 the Great Canadian Oil Sands Plant began separating oil from the sand.

Even conventional oil is a cantankerous resource. It is not edible, does not burn cleanly, is not easy to transport and is potentially deadly. High spirited and independent, it has a mind of its own.

To overcome these stubborn characteristics, early oilmen needed many talents. It was not enough to find petroleum. Indeed, in southwestern Ontario, many old-timers could point you to the nuisance. According to Victor Ross, an early writer about the Canadian oil patch, oil was common.

In 1830 the settlers in the vicinity of Enniskillen, in Lambton County — the extreme western part of Old On-

Ells River, packing dogs with tar sands, c. 1925
(Provincial Archives of Alberta/A5560)

tario – noticed the presence of oil in the swamps of that region. It was called gum oil and was present in such quantities as to be regarded as a nuisance because it killed vegetation and so detracted from the value of the land.

Buying a piece of tainted land that smelled of sulphur was easy. Finding a market for oil was harder. With no cars on the roads and trains burning wood and coal in their steam engines, there was little industrial demand for petroleum products. Consumers rejected the black substance as it came from the ground and only burned amounts of coal oil in their lamps.

First, oilmen had to extract it from the ground, process and refine it, ship it to markets and create a distribution system that was convenient for consumers. The days of gasoline filling stations on street corners and pipelines moving natural gas into homes were a century or more in the future.

These daunting obstacles did not prevent Charles Nelson Tripp of Woodstock, Ontario from incorporating the first oil company in North America in 1851. The International Mining and Manufacturing Company, with Canadian and American investors, set out to make asphalt from the sticky gum of western Ontario.

In 1856, James Miller Williams bought Tripp's producing properties and formed the Canadian Oil Company. Williams knew he had oil, but a supply of fresh water was a more pressing problem. As Williams' crew dug a water well on the Ontario property, oil greeted the shovel at a depth of 20 me-

tres. When Williams boiled the gum, the distillation process "produced a comparatively light, iridescent liquid." Thus, in 1858, James Williams became the proud owner of the first oil well in North America and he named it "Williams No. 1." Accordingly, the nearby communities of Oil Springs and Petrolia sprang up as Canada's first oil boomtowns. Then, as whalers decimated the source of animals for whale oil, petroleum became valued as a fuel and lubricant.

Problems plagued the early oil patch. Production, shipping, refining and distribution were all in their infancy. One man's story illustrates the many things that could, and did, go wrong. Even success has its drawbacks.

On January 16, 1862, Hugh Nixon Shaw drilled the first Canadian "gusher" using a foot-powered spring pole system that hit oil at 50 metres. As Victor Ross, the oil patch writer, reported:

The oil rushed up Shaw's three-inch hole, filled the four-by-five foot well dug through 50 feet of clay, and overflowed at the surface, a great, black, bubbling and gurgling spring of oil.

The oil spread everywhere, shot up in a gusher about six metres into the air, flowed and poured more than 2 000 barrels per day on the ground.

Of the waste in the area, Professor Alexander Winchell of the University of Michigan wrote in 1870 in his book *Sketches of Creation*:

It floated on the water of Black Creek to the depth of six inches, and formed a film upon the surface of Lake Erie.

Early drilling rigs at Canada's first oil field, Petrolia, Ontario, c. 1860s
(Geological Survey of Canada/Glenbow Archives/NA-302-09)

At length the stream of oil became ignited, and the column of flame raged down the windings of the creek in a style of such fearful grandeur as to admonish the Canadian squatter of the danger, no less than the inutility and wastefulness of his oleaginous pastimes.

Winchell estimated the waste at up to five million barrels and called it a "national fortune, totally wasted." Conservation was still a futuristic idea.

Shaw's career as oilman (and industrial polluter) ended in tragedy on February 11, 1863. His obituary in the Cooksville newspaper read:

His death was occasioned by suffocation from inhaling obnoxious gases while in an oil well, into which he had descended for the purpose of pulling up a piece of gas pipe. Was within about fifteen feet of the surface; was heard to be breathing heavily, when he fell back into the oil, and disappeared.

By the 1870s oil was used for lubrication and lighting and the supply from Shaw's gusher did not fill the demand for petroleum products for long. As of the 1880s the people of Central Canada had to import oil from Ohio.

Out West, the Canadian Pacific Railway was looking for a route to the Pacific Ocean. By the early 1880s, track-laying crews were chugging across the prairies while Major Rogers, the surveyor, hounded his haggard team of guides and packers, searching for a way through steep mountain passes.

Meanwhile, unsung heroes pursued more mundane matters. Steam

Drilling rig at Victoria, NWT (Alberta),
1898
(Geological Survey of Canada/Glenbow
Archives/NA-302-11)

George Dawson, third from left, and Geological Survey of Canada crew, Fort McLeod, NWT (Alberta), 1879
(Geological Survey of Canada/Glenbow Archives/NA-302-7)

engines needed coal and water. Coal mines opened to meet the demand and water well drillers punctured the land with their drills. Imagine the surprise at CPR Siding No. 8, near Medicine Hat when they hit natural gas in 1883. A dream come true! But water fouled with stinking sour gas was worse than useless. It was explosive and corroded the water tanks on the trains. So the railway looked elsewhere for water but remembered the natural gas for a later day.

Others were also combing the west for natural resources. Did you hear the story about the big one that got away? In this case it wasn't a fish story, but an oil well at Pelican Rapids near the geographic centre of Alberta.

The federal government sent out an oil-drilling rig from Toronto to the Athabasca River in 1893. Under the direction of the Geological Survey of Canada, drilling began at various sites along the river. The driller was A.W. Fraser, who had experience in the oil industry in India and Ontario.

In 1897, the cable tool rig began drilling at Pelican Rapids. It found a small supply of heavy oil that burned satisfactorily as fuel for the steam boiler. Natural gas, however, was a real problem. According to Mr. Fraser:

"The flow of gas was so great that a cannon ball could not have been dropped down the pipe . . . . The roar of the gas could be heard for three miles or more . . . . Small nodules of iron pyrites, about the size of a walnut, were blown out of the hole with incredible velocity. They came out like bullets from a rifle. We could not

see them going, but could hear them crack against the top of the derrick . . . The danger to the men was so great that they refused to work . . ."

And so the well blew wild, blasting an estimated 8.5 million cubic feet of gas per day into the air for 21 years. Stan Slipper and Claude Dingman finally capped the boisterous blowout in 1918.

The government drilled other wells along the Athabasca. All flowed gas. By the 1920s various schemes to use the gas were under consideration but none proved economically viable. Besides, Ottawa was really seeking the oil supply that fed the tar sands and not hazardous natural gas.

Another tantalizing teaser trickled up from the ground near the American border far south of the Athabasca River. Natives had long known about oil seepages on Cameron Brook near the Waterton Lakes. Suddenly, in 1902 the townsite of Oil City sprang up, complete with surveyed lots, hotels and a post office.

A crusty Irishman, John George "Kootenai" Brown, served as a packer for Dr. George M. Dawson of the Geological Survey of Canada in the West. Brown had lived in the Crowsnest Pass area since 1862, trading with the Natives. In 1874, Dawson asked him if he knew about any oil seepages in the area. Brown asked the locals.

"Some Stony Indians came to my camp and I mixed up molasses and coal oil and gave it to them to drink, and told them that if they ever found anything which smelled or tasted like

George Dawson's Geological Survey of Canada camp at Cypress Hills, NWT (Saskatchewan), near head of Big Plume Creek, 1883 (Geological Survey of Canada/ Glenbow Archives/NA-302-5)

Lunch stop on the prairies for Geological Survey of Canada crew, 1887 (Geological Survey of Canada/Glenbow Archives/NA-302-8)

that to let me know. Sometime afterward they came back and told me about the seepages along Cameron Brook."

In 1888, William Aldridge arrived from Utah and saw potential to develop the seepages. Using a primitive collection system, he captured the oil in burlap sacks, squeezed it out and boiled it down to a thick liquid. This he sold to local farmers and ranchers as a liniment for animals or as a lubricant at a dollar for four litres. A lighter, clearer product in recycled whiskey bottles from local saloons sold for a dollar a litre. Years later his son Oliver Aldridge said that production from the small operation amounted to between 40 and 160 litres a day. For seven years the father and son team made most of their income from this simple oil production system.

Local interest in the potential of oil continued for decades. In 1900, a Pincher Creek newspaper, the *Rocky Mountain Echoe*, wrote in its first editorial as a newspaper:

It is a well-known fact that in the neighborhood of Kootenay Lake there exists a petroleum bed of large extent. The petroleum in many places oozes to the surface forming a deposit so rapidly that it is only necessary to dig a pit a foot or so below the water level in order to collect the oil which forms a thick layer on the water.

The editor went on to predict an oil boom, and the creation of an oil town overnight.

Our district must supply the foodstuffs and other matters necessary for
the existence of such a community; and a new and increasing market would be opened for our farmers and ranchers. Our country would rapidly fill up with settlers, and Pincher Creek would become a more important centre than it is. The oil is there. It is only a question of striking it.

Sure enough, in 1901, A.P. Patrick, John Lineham and John Leeson formed the Rocky Mountain Development Company and began drilling for oil. A cable tool rig and other drilling equipment, valued at $700, arrived by rail from Petrolia, Ontario. From the nearest railway station at Fort MacLeod, teams and wagons moved the supplies 65 kilometres to Cameron Creek. Alex Calvert of Petrolia was the driller, Jerry McDonell of Calgary dressed the tools and Frank Urnberg of Midnapore worked as driller's helper. Drilling began in November.

Success was immediate, if not long lasting. Oil flowed but problems plagued the well. By 1905, the company had a small refinery working and it produced varying grades of oil and tar.

The *Calgary Herald* newspaper reviewed the success of the oil field on January 26, 1906.

Early this month a big strike was made in the oil fields at Oil City. A well that strikes joy to the heart of the stockholders has been opened. They are now beginning to reap the rewards from the development and faith they had in these southern Alberta oil fields.

Many other people placed their faith and money in a handful of other drilling companies that sought black gold in the Cameron Creek oil field. Although showings of oil kept investors motivated for a few more years, the boom went bust by the end of the decade. Not until technologically advanced geological and geophysical exploration methods arrived in the 1950s did the petroleum buried in the Pincher Creek area yield profits for oil company investors.

The Historic Sites and Monuments Board of Canada recognized the Oil City site as the location of Alberta's first successful oil production facility. A cairn and a few ruins are all that remains of western Canada's first oil boomtown.

So why all the excitement about oil discoveries in Canada? Since the 1880s, suppliers in central Canada had imported oil from the United States. Cheap, reliable and convenient sources of petroleum products could create employment and help bring Canadian industry and society into the 20th century.

As of 1905, there were only 565 automobiles in Canada. In 1908, the first Canadian service station opened at the Imperial Oil warehouse in Vancouver. This station used a garden hose connected to a kitchen water tank filled with gasoline to serve its customers. Luckily cars and trucks were not going to demand a huge supply of gasoline just yet!

But Canadian consumers were falling in love with oil products. Lamp oil was commonly used in many homes. Lubricants came from petroleum. Some people traded in their wood and coal stoves for new models that burned fuel oil. And industry was becoming increasingly interested in petroleum and natural gas to replace sooty coal and wood.

Therefore, in 1904, Parliament passed the *Petroleum Bounty Act*. Ottawa encouraged the search for petroleum with an incentive that began at 1.5 cents per gallon (.375 cents per litre). The government paid out $350 000 in 1904. Over the years it amended the Act several times and the bounty declined to $80 000 in 1924, the year Ottawa ended the payments. For those two decades the central government directly encouraged resource exploration.

In spite of the incentives, most discoveries were accidental. Based on samples from coal outcroppings, CPR employees drilled for the elusive black mineral near Medicine Hat in 1890. Gas greeted their drill, just as it had in 1883 when the CPR drilled for water. Medicine Hat politicians approached the railway company and arranged for further drilling to ensure a reliable supply of natural gas.

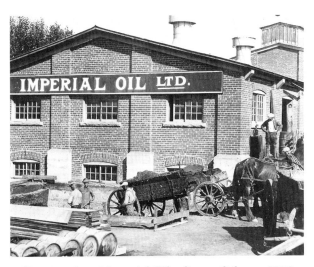

*Construction at Imperial Oil refinery, Calgary, 1928
(B.L. Nowers/Glenbow Archives/NA2849-102)*

Soon the community boasted all the gas it could use in its homes and industries. Gas street lights burned night and day. On a visit to Medicine Hat in 1907, Rudyard Kipling fell in love with the prairie town. Not only was it blessed to have this fantastic supply of natural gas, he also thought it entertaining to have "all Hell for a basement."

In spite of this resource, Medicine Hat did not become the centre of the petroleum industry in Canada. Operating decades before pipelines spanned the continent, Medicine Hat gas could only supply the local market. With no regional, national or international transportation system to deliver the valuable commodity to consumers elsewhere, Medicine Hat could only entice potential customers to relocate to the prairie community.

Archibald Wayne Dingman, early 1930s
(E.B. Curlette/Glenbow Archives/NA1424-1)

In the city of Calgary, one natural gas well of modest production proved successful. In 1908, the Calgary Natural Gas Company, organized by A.W. Dingman in 1905, found gas at its well on the Colonel Walker Estate in east Calgary. Due to its proximity to industrial and consumer markets, this well helped Calgary become the centre of the Canadian petroleum industry.

In February, 1909, Eugene Coste's "Old Glory" well at Bow Island struck a flow of eight million cubic feet of gas per day. Although almost as far removed from a major metropolitan centre as was Medicine Hat, Bow Island had Coste on its side. He set up the Canadian Western Natural Gas, Light, Heat and Power Company in 1911 and purchased the rights to supply Calgary with gas.

Early the next year, pipeline crews began building the world's longest large diameter gas pipeline (40 centimetres) from Bow Island to Calgary. After only 86 days of work, the 275-kilometre line was in the ground, serving customers at communities along the line and at its terminus in Calgary.

On July 17, 1912, gas arrived under its own pressure in Calgary. A celebratory flare in downtown Calgary attracted 12 000 spectators to the opening ceremonies. Most people remembered the spectacle, but few recalled the problems. Mr. P.D. Mellon, supervisor for the pipeline at the time, recalled an explosion in the pipeline that almost squelched the celebration.

"Early that afternoon the line blew up in a slough near DeWinton. Things looked pretty gloomy but we rushed several gangs of men down there and we were able to get the line coupled up again and the pressure built up.

That evening the celebration went ahead as planned.

"Eugene Coste and his wife were there and Whitey Foster was in charge of the valve control. At a signal from Mr. Coste, Whitey turned on the valve and he turned it on plenty, because coming out of this standpipe there was a tremendous amount of dust, then stones and great big boulders, two or three pairs of overalls, pieces of skids — almost everything came out."

Mrs. Coste lit the gas column with a Roman candle.

"Then Mr. Coste signalled to turn it down. Whitey thought he meant to turn her on more. Whitey opened her up again and this almost caused a panic. People were backing up into each other and yelling at this terrible flare going up into the air.

On July 24, gas began flowing to customers. The *Calgary Herald* stated:

This morning marked a new epoch in the history of lighting and heating insofar as gas is concerned in Calgary. Artificial gas in the city is now a thing of the past. The ringing out of the old and ringing in of the new has

taken place, and the natural product has supplanted the artificial.

For a while, it seemed like Calgary's fuel problems were over. Eventually, even the enormous gas reserves at the Bow Island field proved insufficient to the city's needs.

Meanwhile, at the other end of the country, in New Brunswick, petroleum discoveries suffered similar problems. In 1908, the Stony Creek oil field opened for business. Oil was known long before the 1900s in this part of Canada, but the early discoveries amounted to little or nothing. In 1859, H.C. Tweedle found oil and gas in the Dover oil field in New Brunswick.

For decades, the Maritimes had relied on wood and coal for fuel. Slowly they began converting to cheap oil, from the Pennsylvania and Ontario fields. But supplies were unreliable and prices fluctuated from $2 per barrel in 1862 to $8 in 1864. This tumult encouraged exploration elsewhere.

Fear of competition by foreign companies overseas caused American and British companies to seek oil fields to control outside their borders. As a result, New Brunswick's enticing ownership of the mineral rights and generous 99-year leases attracted developers to the province. The New Brunswick Petroleum Company Ltd., a Scottish company, formed in 1893 and drilled enough to qualify for the 99-year lease. It was successful and eventually produced commercial amounts of oil and gas and supplies Moncton to this day. Only by opening up exploration to outsiders and rewarding them with attractive leases was this province able to lure the investors who could develop the oil fields.

Back on the prairies, the long cold winter of 1914 proved too much for even the massive Bow Island gas supply. Small amounts of additional gas from the Chin Coulee and Foremost fields helped with winter demands. Calgary and southern Alberta desperately needed a new source of reliable natural gas. As it turned out, that supply was on Calgary's doorstep, at Turner Valley.

The dormant oil field on the outskirts of Calgary was not a typical minor gas play. For more than a generation, Turner Valley was the largest, most famous, most spectacular and unruly producer in all of Canada and the whole British Commonwealth. Decades later the giant discoveries of the 1940s and 1950s dwarfed this early producer, but in the days before continental pipelines, Turner Valley was the only game in town.

The story of the discovery of Turner Valley is not without intrigue. Legends grow bigger each passing year. We will never know for sure who first noticed the gas and oil seeping from the ground along the creek bank.

A Stoney elder at the Eden Valley Reserve west of Longview, Johnny Lefthand, tells the story of Anna Dixon. According to him, Anna discovered the oil and used it for medicine. One day she showed it to the whites. They gave her food and trade goods as payment for helping them discover the oil seeping out of the banks of Sheep Creek.

Another tale says that Sam Howe and Negro John Ware, early cattlemen in the area, crossed Sheep Creek in 1888 and stopped to water their mounts. When their horses balked at the water, Ware examined it more carefully and found it covered with a slick of oil and smelling of sulphur. He lit a match and watched with surprise as the water caught fire. After it blazed out, they bottled some oil and sent it away for analysis. Coal oil from the coal beds, stated the experts, and that made sense to the locals.

Little seepages became well-known along Sheep Creek but not until 1911 did anyone take them seriously. William S. Herron captured a gas sample, as the story goes, and sent it off to the University of California. The tests came back positive for wet gas so Herron began buying up property and setting up an oil company.

Raising money for a speculative oil venture was tough work, so Herron pulled out all the stops when he brought driller A.W. Dingman and lawyer R.B. Bennett out for a visit. Using a fishing trip as a pretence, he got them to Sheep Creek. Then he lit one of the natural gas seepages and cooked lunch for his impressed guests. This story may or may not also be true.

Finally, in July 1912, Herron, Dingman, Bennett and others joined forces and formed the Calgary Petroleum Products Company. Drilling began in early 1913 and the usual financial problems plagued the young company. By October, the well hit enough pockets of gas that speculators took out leases on more than 150 square kilometres of land in the area.

Rumors and stories continued through the winter until May 14, 1914. At a depth of 824 metres Calgary Petroleum Products No. 1 well came in with a roar. Locally called the "Dingman No. 1" well after the head driller, it attracted attention from all over the country.

A light-colored oil accompanied the four million cubic feet of gas per day that flowed from the well. One report called it "so pure it was used as car fuel at the well site." Calgarians flooded out of the city in every available car to visit the famous oil field.

There the inquisitive crowd of embryo oil magnates amused itself by filling bottles with the gasoline-like liquid which had been taken from the well, and in surmising what Dingman Well No. 2 (located a thousand feet from No. 1) might be worth.

The boom that followed staggered the imagination. Every nook and cranny in downtown Calgary became an office for some oil company. Likely as not, the company had no land, drilling rig or even a board of directors. Most just printed up stock, traded it for the cheques and cash of the gullible and excited public, and got out of town on the next train.

One old woman demanded that a railway employee sell her some oil stock. Pressed into service alongside the other cashiers where the business of selling train tickets was second to selling stock, he tried to warn her. He said he knew nothing about the company or its reputation. "Oh, that doesn't matter," was her reply, "anything will do as long as I get some stock."

After the crazed investment in June and July of 1914 due to the Dingman #1 discovery, most of the people who invested money in the Alberta oil

fields had only worthless stock to show for their efforts. Consequently, the Alberta government appointed a Utilities Commission to "prevent such reckless exploitation as occurred during the late spring and summer of 1914." Litigation resulted in at least one man spending time in jail for fraud.

Although many people fell for the scams, the Turner Valley oil field itself was real enough. More substantial discoveries in 1924 and 1936 eventually proved the true value of the field, but the wet gas and oil in this field put Calgary on the map as Western Canada's oil centre.

As war loomed in Europe, the interest in Turner Valley faded. But drilling continued and Calgary Petroleum Products built the first gas-processing plant in Canada. On December

31, 1921, Turner Valley began supplying the pipeline that serviced Calgary with natural gas. Close to a ready market, the future looked bright for this oil field.

The final major oil discovery in early Canadian petroleum history came in the North. In 1917, Victor Ross wrote:

In the great Northwest territory — most of it beyond the reach of economic development at present — occur some of the most promising indications of oil fields in Canada.

His words are almost as true at the end of the century as they were at the beginning.

Various traders, explorers and other white interlopers noticed oil in the northern part of Canada. Finally, in 1914, the Imperial Oil Company Limited of Toronto, a subsidiary of Standard Oil of New Jersey, moved in to explore for oil in a serious manner.

Using small river steamers and scows, Dr. T.O. Bosworth and his crews explored the Mackenzie River system, looking for the oil field that would open the North. Where even these craft were too large, they took to canoes and sometimes even plodded across the soggy muskeg on foot.

By summer's end Bosworth's survey was completed and he summarized his observations for the company. Seven locations might produce oil, he warned, and Imperial should gain control over a large portion of land in the area to prevent others from staking claims.

Once the fields came into production, he remarked, it would require substantial support systems to make them economically viable. Access to the remote area was most problematic.

Fort Norman gusher, NWT, Ch. Taylor turning on well, 1920
(Provincial Archives of Alberta/B1062B)

First four men to reach Fort Norman, NWT by airplane. Left Peace River, Alberta on May 29 and arrived at Fort Norman June 12, 1921. Total flying time of 12.5 hours. W. Hill, mechanic; Ted A. Link, geologist; Elmer Fullterton, pilot; W. H. Waddell, surveyor and navigator. (Ted Link.Glenbow Archives/NA-4552-1)

The all-water route on the Mackenzie River and the Arctic Ocean was tempting, but the short summer on the far northern sea made it unreliable. A complex system of river travel, portages around rapids and a railway linkage to Fort McMurray were also inconvenient but proved the only viable route.

War intervened and it was not until 1919 that Imperial appointed Dr. Ted Link chief of exploration for the Northwest Company in the region north of Fort Simpson. Link immediately decided to drill on a location first chosen by Bosworth in 1914. It was also near a good river landing. Justifying his choice, he wrote:

The location of the prospect hole has

been made right in the centre of the seepage area, and its seems almost certain that at least a good show of oil will be encountered at a depth of 30 to 100 ft. If oil is struck at a greater depth, more than a good show is expected.

During the winter, Link's men stayed behind at the well site. Expecting to shoot game for meat, the crew ran short and had to shoot the ox, Old Nig, for a supply of meat. Finally, in July, 1920, Link arrived with supplies and additional men. Spirits rose.

The new drilling crew helped the work along and in August the well started to show oil. On Tuesday, August 24th, Alex Patrick, son of famous driller A.P. Patrick, wrote in the log:

Ran bailer in the hole; after running bailer twice, the irritation started the well flowing; this time there was nothing to restrict the flow of oil; which rose to a height of seventy-five feet above the derrick floor.

And so Imperial had its first oil gusher in the North, many hundreds of kilometres from the nearest railway, pipeline or markets.

When he arrived back at Edmonton that fall Link told the Edmonton Journal on October 16, 1920: "This is the greatest oil field in the world – stretching to the Arctic . . ." Although the secret was out, the remoteness of the area prevented its exploitation and protected it from most outside development forces until the emergency conditions of World War II created a panic for Northern oil.

In most of Canada, as in the far North, shows of oil were less valuable than fur pelts in the early days. Canadian society first had to change from dependence on wood and coal to oil and gas. Only then did development of early petroleum discoveries seem possible. Many obstacles still stood in the way of developing oil and gas. However, as technology advanced, and resources and markets became more accessible to each other, the Canadian petroleum industry became a central part of the economy.

Resource Rich

. . . developing Alberta's Inheritance

Spectacular gas flare February 13, 1947 at Imperial Leduc No. 1
(Harry Pollard/Provincial Archives of Alberta/P2722)

I watched Thursday afternoon while IMPERIAL-LEDUC No. 1 kicked itself off and staged a spectacular 'flare' scene. In the small hours of this morning I shivered in a raw wind while my hand on the flow pipe recorded the steady pulsating of oil heading for the storage tanks and gas heading for the flare. . . .

In his *Daily Oil Bulletin* the undisputed dean of oil patch journalism, Carl O. Nickle, sang the praises of a spectacular oil discovery near Edmonton in 1947. As a tireless booster of the Alberta oil industry in the form of writer, member of parliament and investor, he was in the crowd witnessing the sweetheart of a find on February 14 that catapulted Alberta from a regional oil producer into a continental oil storehouse.

Oil did not appear miraculously at Leduc that cold winter day. Alberta's buried treasure lay hidden deep below the plains, where Nature had tucked it into the rock formations of the foothills for centuries. Ranching and agriculture alone could not make Alberta wealthy and powerful: the natural resources trapped in ancient geological formations required liberation.

First someone had to notice the oil seeps. Others had to imagine a society based on petroleum. Still others had to create the technology to seek, find, produce, refine and distribute oil and gas. And society had to change. It had to embrace petroleum as a friend. In short, Canadians had to fall in love with oil before Albertans could begin to cash in on their inheritance.

Petroleum is not like other natural resources. The first Europeans exploited the fishery along the eastern coast of the North American continent because the resource was so obvious. Natives trading plentiful fur came next. Then the Europeans harvested massive trees for timber. As they moved west, they took advantage of the great plains to grow grain for world markets. They mined coal, gold and silver from the vast storehouse of treasures that awaited the prospector in the Canadian frontier.

Oil is different. It is not an obvious resource like wood or wheat. It is not as easy to mine as coal. In fact, finding it in commercial quantities is tricky. One well can reveal nothing. The next might be a gusher. Good fortune is as important to an oil company as all the expertise in the world.

Albertans are extremely lucky in that the Western Canadian sedimentary basin contains oil and gas. Stretching from the great plains north to the Arctic Ocean, this geological formation contains vast deposits of petroleum.

No one is quite sure how oil and gas formed beneath the earth's crust. Inorganic theories suggest that some force, perhaps volcanic, created petroleum deep inside the earth. The organic explanation is more widely believed, but still not well understood.

According to this theory, decomposing plants and animals created coal, oil and gas. Trapped under rock formations, they awaited discovery many thousands of years later. Regardless of the explanation, petroleum lies buried under many parts of Canada. It is Canadian's responsibility to develop this bountiful resource wisely.

In 1858, the North American's initial oil find made Petrolia, Ontario, famous as Canada's first oil field. Although spectacular, Petrolia's production failed to meet the explosive consumer demand and by the 1880s Canada had to import oil.

In the 1920s, Alberta surpassed Ontario as Canada's primary petroleum province. Even though Alberta produced only 884 barrels of oil in 1924 compared to Ontario's 154 368, in 1925 Alberta's share grew to 67 percent of the total, or 183 491 barrels compared to Ontario's 143 134. In the years that followed, Ontario's production remained stable while Alberta shattered all national records. Production in 1930 was 1 396 160 barrels, 92 percent of national production. In 1940, 10 117 078 barrels, or 98 percent of Canada's oil came from Alberta.

Other statistics also show Alberta's significance as an oil centre. During 1925 the number of oil field employees in Alberta exceeded those in Ontario. By 1926, Alberta boasted six refineries and 33 companies drilling for oil, most with their offices in Calgary. Twenty-two gas producers worked fields at Wainwright-Fabyan, Redcliff, Viking, Wetaskiwin, Medicine Hat, Bassano, Suffield, Foremost, Bow Island and Turner Valley. In 1927, $20 426 000 or 90 percent of capital employed in oil production in Canada was at work in Alberta.

Obviously, Canadian oil companies believed that Alberta was the place to look for oil. But the search for oil was more scientific than just drilling at seeps. Increasingly, highly-educated experts put their considerable skills to work in the quest for petroleum.

British Petroleum Ltd. Well No. 3 at Wainwright, July 18, 1925
(Carsell/Provincial Archives of Alberta/A10,793)

Royalite employees at Turner Valley, March, 1926

Geologists were among the first to use their education to find oil. The Geological Survey of Canada was an early source of information on the general geology of the West. Later, in 1912, Dr. John. A. Allan established the geology department at the University of Alberta. He produced the first geological map of Alberta in 1920.

But hard rock geologists considered petroleum geologists as poor cousins. And they were. Gold, silver and precious minerals attracted the glamor. Many geologists looked down their noses at their counterparts who sought smelly gas and sticky oil. Still, in 1925, the Canadian Western Natural Gas Company established a geological department at its Calgary office.

A few independent geologists also worked in Calgary during the 1920s. These specialists periodically lunched together and exchanged ideas. In 1924, oil fever hit again when a deeper discovery tapped a more plentiful source of wet gas at Turner Valley.

Once again, charlatans bilked the public of its money by posing as oil experts. In response, the reputable geologists began working on a plan to protect their reputations.

Organizations with an interest in the petroleum industry were not new. The Canadian Institute of Mining and Metallurgy (CIMM) sprang up in 1898. In 1917, American geologists formed the American Association of Petroleum Geologists (AAPG). Producers formed the Ontario Natural Gas Association in 1919. The Association of Professional Engineers, Geologists and Geophysicists of Alberta (APPEGA) came together in 1920.

In 1926, Alberta producers formed the Oil Operators' Association of Alberta, later known as the Canadian Petroleum Association and now called the Canadian Association of Petroleum Producers (CAPP). Also in 1926, CIMM formed a natural gas division to meet the needs of a growing industry. Finally, in 1927, the time was right for a

(Kilroe/Provincial Archives of Alberta/84.54/1)

recognized society of petroleum geologists in Canada. Called the Alberta Society of Petroleum Geologists, it changed its name to the Canadian Society of Petroleum Geologists in 1973.

Dr. Ted Link was an early leader in the Canadian geological community, and certainly the most famous. Born in Indiana and educated at the University of Chicago, Dr. Link worked for Imperial Oil in the West in the early years. Instrumental in finding oil at Norman Wells, he was also an important part of the effort to produce oil for the war effort during World War II.

Dr. Link once ordered shotguns for his geological field crews working in the North. After the guns arrived without shells, he repeatedly ordered ammunition, without success. Finally he wired the U.S. commanding officer at Whitehorse, saying, "A bear ate two of my geologists yesterday; now will you send the god damned shells?" They arrived the next day and the search for oil continued. Although oil

companies invested considerable capital in Canada between the two world wars, they found no new oil fields. However, the events of these two decades prepared the economic, social and political climate for the developments that followed World War II.

In 1930, for example, Ottawa transferred control over natural resources to Alberta and Saskatchewan. Before that date, all income from oil and gas belonged to the federal government even though Ottawa created these provinces in 1905. This transfer gave them the potential to reap great rewards if the natural resources proved plentiful.

Six years later, the Turner Valley Royalties No. 1 well ushered in the era of crude oil in Alberta. On June 16, 1936, the well began flowing 850 barrels of crude oil per day, proving the Turner Valley gas field also included a valuable layer of oil below the gas cap. The discovery created another drilling boom in the old oil field. Millions of

Flare at well in the Turner Valley oil field, c. 1930s
(Glenbow Archives/NA67-143)

Welding pipe for line from Turner Valley to Calgary, c. 1930
(Harry Pollard/Provincial Archives of Alberta/P1973)

Lowering pipe into ditch for line from Turner Valley to Calgary, c. 1930
(Harry Pollard/Provincial Archives of Alberta/P1972)

dollars flowed into the provincial economy and investors smiled all the way to the bank. During an economic depression that crippled the rest of the country, petroleum wealth made Alberta look charmed.

By the end of 1938, Carl Nickle's *Daily Oil Bulletin (DOB)* announced the extent of the good news. Alberta's oil production doubled in 1938, to 6.5 million barrels of crude oil. Ironically, while Canada was importing 30 million barrels to meet consumer demand, Alberta oil wells were limited by production quotas. The catch was the transportation system. There was no way to get Alberta oil to other Canadian consumers. In his year-end editorial, Nickle begged:

Wake up, Canada! is the plea of Alberta Oil to the nation. Open your markets in this New Year. Encourage an industry that can pave the way to Security for the Empire, help balance budgets, pull your railroads out of the red, and add hundreds of millions of dollars to the national pocketbook. Look at the record, then try to slumber on!

Although the oil boom helped solve Alberta's financial problems, it also created headaches. On the positive side, the new Social Credit government reaped untold wealth from the new discoveries. Much to its chagrin, it regularly failed to balance its budget, constantly underestimating its natural resource revenues.

But there were also more disturbing problems. The resurgent, heady, wildcatting and lucrative petroleum sector was also greedy, ill-mannered, unregulated and wasteful. It was an example of the best and worst characteristics of a boomtime economy.

Many factors contributed to the chaos. Demand for oil and gas was high. Everyone was looking for an end to the depression that had robbed the society of hope. Most of all, a lack of effective government regulation and access to markets promoted profligate waste of the very natural resource that promised to make Alberta rich and powerful.

For example, most producers had no market for the gas they had to produce to separate the liquids out of the wet gas. The Royalite Oil Company, a subsidiary of Imperial Oil, had a monopoly on the sale of Turner Valley gas to Canadian Western Natural Gas and the only pipeline to Calgary. As a result, independent producers could only sell the liquids (similar to gasoline). The remaining product, natural gas, was waste to them. They saw no alternative but to flare it. According to one estimate, Turner Valley producers burned off $10.00 of gas to produce $1.00 worth of oil.

Regardless of the apparent economic necessity of flaring "excess" natural gas, this practice encouraged wasteful behavior among the producers. They flared the gas as quickly as they could to sell the liquids. Greed

ruled.

This selfishness was not only unethical, but a flaw that some might condone in the midst of a booming oil field. The government stepped in to regulate the industry because these greedy production practices threatened to reduce the ultimate amount of recoverable oil – and that is exactly what happened.

When industry proved unwilling to solve the problem, government took measures to control the waste. The oil companies were not amused. In reaction to the government legislation of 1931 and 1932, producers used their considerable economic clout to hire lawyers to tie up the government's conservation decisions in the courts for years. Although Alberta took control over its natural resources in 1930, it was not until 1938 that it finally gained the legal power to control waste in the Turner Valley field.

Finally, on February 26, 1938 the Alberta Petroleum and Natural Gas Conservation Board received the power to regulate the petroleum industry in the province. Its duties included limiting the waste of gas in the Turner Valley field and managing efficient development of the oil field. In retrospect, the Conservation Board's mandate was narrow: maximize potential economic yield from the reservoir. In later years the government gave the board the power to investigate and regulate environmental and social aspects of the development process.

But government regulators could not solve the most vexing problem for western Canadian producers – the lack of markets for oil and gas. In

Geological survey crew at Threepoint Creek, Alberta, December 2, 1941. Glenn Fox, student assistant; Bob Sharp, packer; Jack Sinclair, cook; Con Hage, in charge. (Glenbow Archives/NA-3834-3)

S.J. Rosel's truck unloading valve in the Turner Valley oil field, c. 1943 (Harry Pollard/Provincial Archives of Alberta/P1824)

Rig crew working without hardhats on a Turner Valley rotary drill rig, c. 1940s
(Harry Pollard/Provincial Archives of Alberta/P1238)

1939, independent oil operators lobbied both the Alberta government and British financiers to build a pipeline from Turner Valley to the Great Lakes. The Canadian branches of international oil companies, however, did not support the project. Their companies supplied the Eastern seaboard and the Great Lakes communities with cheap oil from their offshore operations and therefore saw little need to build an expensive line across the continent. The independents, with limited regional markets, cried "foul!"

Just in time, war broke out. Although it did not solve the pipeline question, it created new regional markets for western Canadian oil. In September, 1939, oil men started enlisting for service in World War II. Who better

to volunteer for this conflict than the men who built the main support system that would make it possible to win the international battles?

Although troops fought World War I in the trenches, World War II relied on the internal combustion engines in jeeps, trucks, airplanes, ships and submarines – all fueled by oil.

As early as the 1920s, prophets saw the potential of petroleum in the international arena. On December 28, 1929 the *Western Examiner*, a Calgary oil industry newspaper, noted:

Possession of Petroleum Deciding Factor in War: . . . Petroleum will largely govern the fortunes of nations in the years to come. Already it is the cause of one of the greatest fights for possession that the world has ever witnessed. Armies, navies, money, even population, will count as nothing against it . . . . We are all for peace; we want no more of that horror that ended now eleven years ago. But, if war should ever again be visited upon an unhappy world, it will be decided by petroleum. The present effort to control supplies is more gripping than any drama ever written.

So when war broke out in Europe, the shut-in production in Turner Valley had a market. Not that Alberta oil went directly into the machinery that fought the war in France. Instead, petroleum products from Canada's largest oil field went into the engines that powered the vehicles used in training Allied troops.

Although the Alberta government rationed the development of the Turner Valley field before the war, during World War II Ottawa stepped in and forced companies to open the taps on the oil wells and produce at maximum capacity. Under wartime emergency conditions, Turner Valley production practices once again endangered the long-term economics of the oil field.

As a result of the war, in 1940 Home Oil drilled exploratory wells in a new area in the north end of Turner Valley field. The drill bit into success and, with an almost unlimited market for the product, oil flowed like never before. That same year, the British American refinery in Calgary announced plans to produce aviation gasoline for pilot training.

Decades later, it is hard to recreate the emotion generated by wartime fears or fully comprehend the important supporting role Alberta oil played in the conflict. But the March 20, 1942, issue of the *DOB* proclaimed:

ALBERTA'S OIL FIELDS ASSUME NEW IMPORTANCE AS ALASKA DEFENSE HIGHWAY CONSTRUCTION STARTS!

The United States Government has called Turner Valley into action against the Japs. The Valley has been called upon for the many thousands of barrels of high grade aviation gasoline and motor fuel needed for the construction, maintenance and protection of the new vital defence highway to Alaska.

Signs reminded refinery workers of the strategic importance of the Royalite Refinery at Turner Valley to the war effort, February 7, 1945. (Harry Pollard/Provincial Archives of Alberta/P2384)

During 1942, Alberta oil production passed the 10 million barrel-per-year mark for the first time in history, setting a record in production for the seventh consecutive year. On December 31, 1942, the *DOB* reported:

The great British Commonwealth Air Training Plan continued to expand during 1942, and by year's end its planes were consuming two million gallons of hi-test aviation gasoline per day, a large part of it over the Plains of Western Canada. Much of the 87-octane fuel is being provided from Alberta oil. At year's end, Western refineries were ready to commence production of fighting grade 100-octane fuel from the high grade of Turner Valley.

On April 9, 1943, Ottawa created Wartime Oils to encourage exploration and development in Turner Valley and to thus help the war effort. This precursor to later federal incentive programs lent money for exploratory wells. Loans were due, with interest, only after successful completion of the oil wells. Most oil companies jumped at the chance to carry out exploration using government finances.

In spite of a massive infusion of government capital and large expenditures by Canadian branches of multinational corporations, the middle years of the 1940s brought disappointment.

Tempting shows at wells in the plains during 1943 suggested future prospects. Wells in the Vermilion area and others throughout Alberta teased. Although some found shows of oil or gas, they failed to make the Big Find.

Then, at a foothills well west of Calgary, Shell Oil of Canada made an important discovery at Jumping Pound in 1944. Hope sprang anew from the hearts of investors who expected another massive oil discovery similar to the Turner Valley field. A valuable gas discovery it was, but it did not produce the long awaited new oil field in Alberta.

At this point, near the end of the war, Alberta's oil production was dropping. Turner Valley gave of its wealth at maximum capacity, but the effort reduced the oil field's potential. Production peaked in 1942. When Jumping Pound turned out to be only a gas field, the search for oil returned to the plains. By this time, however, a new breed of scientist helped in the search.

Geophysics in the oil patch arrived in Canada in the 1930s as a science still in its infancy. As a result, geologists scoffed at the "jughounds" and "doodlebuggers." Not many years or much technology separated the geophysicist from the dowser and his forked stick, wandering across the plains, claiming to find oil or water based on the twitching tool.

However, prominent geologist Charles Gould of the Oklahoma Geological Survey gave his qualified support to geophysical methods in an article in the *Calgary Oil and Financial Review* dated March 21, 1931. He was convinced that geophysics gave oil companies a 50:1 improved chance of finding oil compared with wildcat drilling alone. But, he warned that even though oil companies had spent forty million dollars on geophysical research in the previous decade, the "drill is actually the only oil finder."

Floyd "Flip" Brown talks on the telephone with his seismic party chief, Alberta, 1940s (Imperial Oil Ltd./9799)

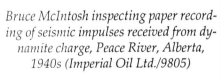

Bruce McIntosh inspecting paper recording of seismic impulses received from dynamite charge, Peace River, Alberta, 1940s (Imperial Oil Ltd./9805)

Seisomgraph rig and derrick preparing shot holes for dynamite charge in northern Alberta, n.d. (Harry Pollard/Provincial Archives of Alberta/P1328)

Jug hustler setting jug to receive impulse from dynamite charge, Alberta, 1940s (Imperial Oil Limited/9383)

By the mid-1940s, geologists and geophysicists were working together. Seeking oil in the Canadian West, they found nothing new. Based on their knowledge, Imperial Oil's geologists and geophysicists decided to drill in the Edmonton area. Geologist Doug Layer recalled:

"It should be realized that our seismic techniques in 1946 were quite primitive when compared to today's advanced technology . . . The seismic system consisted of 12 amplifiers, recording from 12 geophones spread over a quarter mile, on photographic paper. The energy source was 40–60 pounds of dynamite shot in 60 foot holes."

Imperial's seismic survey showed promise of a formation that might hold oil. Therefore, Imperial drilled its first well at Leduc in 1946 and the events of February, 1947 proved the survey speculations right. Many of the Leduc old-timers are still alive. For many Albertans and Canadians, the name "Leduc" is synonymous with Alberta's modern oil industry.

This oil field justified building pipelines east, west and south through which prairie oil and gas flowed to distant markets. Indeed, the spectacular media event created to usher in the new oil field on that cold Thursday in February was calculated to make a big impression on the province and the nation.

*Spectators flocked to see the black gold on February 13, 1947 at Imperial Leduc No. 1
(Harry Pollard/Provincial Archives of Alberta/P2720)*

The lake of oil at Atlantic No. 3 in 1948 threatened to erupt into an inferno. It bubbled and fumed for months before pipelines pumped it away.
(Harry Pollard/Provincial Archives of Alberta/P2824)

But even a carefully orchestrated publicity stunt relied on nature. Imperial Oil invited politicians and the media to the site to witness the coming in of the well and a new oil age. According to Vern Hunter, the driller on that famous well, not all went as planned. People showed up that morning, as requested, but the technology did not cooperate. The swabbing unit broke down and delayed the momentous occasion until late in the afternoon.

By four o'clock the less hardy had shivered their way back to town, but the faithful saw a beautiful ring of black smoke go floating skyward – a good omen for the oil industry in Western Canada . . . . Then with a roar the well came in, flowing into the sump near the rig. We switched it to the flare line, lit the fire and the most beautiful smoke ring you ever saw went floating skywards.

That smoke ring signalled a turn in the fortunes of Alberta's oil industry. Before that ring, there was only the failing hope that another oil field like Turner Valley might exist along the foothills. After that smoking signal, Imperial Leduc No. 1 was the proof of oil in a previously untapped geological formation, on the prairies. Explorationists in the oil business began applying new theories to their understanding of oil reserves. Crude, black oil underlay large parts of Alberta. All the oil companies had to do was drill into the vast underground reservoirs.

On December 23, 1947, Carl Nickle wrote in the *DOB*:

In the long history of the Canadian oil industry, the brightest chapter written in the shortest time is headed "Leduc." Written in the ten months since February, 1947, it is the beginning of a series of chapters yet to be written, for time still hides much of the Leduc story, and the implications of the "Leduc Discovery" may well bring new discoveries which might at long last make Canadians self-sufficient in petroleum.

In 1948, the Leduc discovery well and its quickly drilled neighbors produced enough oil to push Alberta once again over the 10 million barrel production milestone for the year. Turner Valley peaked at that level in 1942, but by 1948 was only producing about half the oil attained during the wartime emergency. Therefore, the discovery of oil in the plains ensured Alberta's continuing position as Canada's most important oil producer.

With success also came excess. Only months after the exhilaration of the Leduc discovery, a nearby well blew wild, spewing hundreds of thousands of barrels of oil on a farmer's field. Western Canada's most spectacular blowout is also an important part of Alberta's oil patch story.

Drilling with inadequate protection against a blowout, the Atlantic No. 3 well got into trouble in early 1948. The gusher literally spread oil throughout the area. When the wind was just right, oil settled on clotheslines in the nearby town of Devon. Berms built to hold the lake of oil leaked and oil flowed into the North Saskatchewan River and down to Edmonton.

Early attempts to plug the well

with cement were fruitless. One man said, "You might as well have pumped it into the North Saskatchewan River for all the good it did." So they tried to seal the well by pumping it full of a variety of products. The many tons of unsuccessful materials included chicken feathers from Montreal, sawdust, mud, lime, cottonseed hulls, redwood fibre, insulation, grain, sugarbeet pulp – almost anything they could find. Nothing worked.

And still the lake of oil grew. One oilman recalled the scenery at Atlantic No. 3.

"The whole quarter was what Hell must have looked like – a field of surging, bubbling oil, water and mud."

By May, 1948, the Conservation Board had a real emergency on its hands. The waste problem at Turner Valley was immense, but it had not endangered lives. By contrast, the blowout at Atlantic No. 3 spewed oil into the air, got into Edmonton's water supply, and no one seemed to have the ability to control the wild well or clean up the massive spill.

Then the conservative Social Credit government took a drastic move and the Conservation Board took over the well. Expropriation of private property is a drastic action for any government, but the wild well was an unusual situation and called for immediate and decisive action.

Efforts to kill the wild well continued for months. On Monday, September 6, 1948, the rig toppled. A few hours later, the well caught fire. Tom Wark, an experienced Turner Valley driller, watched the well ignite.

"A ball of fire danced over the derrick floor about fifteen to thirty seconds and then the flame crawled down the side, into the main crater around the well. It never really exploded. It then just seemed to start everything on fire at once."

The well burned for 59 hours, creating what Carl Nickle called a Funeral Pyre that sent flames shooting 250 metres into the air and smoke billowing up two kilometres. *A long bank of smoke spread for 100 miles across the Central Alberta Plains,* wrote Nickle. The Conservation Board oversaw the drilling of two relief wells. One pumped water into the wild well and snuffed out the flames. The other capped the runaway with cement.

After the excitement died down, government engineers estimated the spill at 1 407 979 barrels of oil and about six percent of the gas in the oil field. The loss of gas decreased the eventual recovery of oil from the field by another 2.5 million barrels, or nearly $8 million based on the price of oil in 1948.

In retrospect, good fortune smiled on the men who encountered the power unleashed at that well. They were lucky that lack of safety equipment did not lead to deaths or serious injury. Fortune also prevented the whole lake of oil from catching fire.

Atlantic No. 3 caught the Leduc oilmen flat-footed. They had not suspected the great underground pressure that lead to the blowout. They were inadequately prepared to deal with the wild well.

The Alberta government had to step in and take over the problem. Only the work and intelligence of a group of dedicated and lucky men finally extinguished the fire and plugged the well.

According to the late Les Rowland, editor of *Oilweek Magazine* for many years, the blowout at Atlantic No. 3 was "more of a sensation than a problem." The landowner got generous voluntary financial compensation and no one really bothered to investigate the effects of the oil on the farmland.

No one had time. The major discoveries came one after another for more than a decade. Vast amounts of crude oil greeted the drill at Redwater in 1948, at Golden Spike in 1949, at Fenn-Big Valley in 1950, at Wizard Lake in 1951, at Pembina in 1953, at Swan Hills and Virginia Hills in 1957 and at Judy Creek in 1959. Sandwiched between the oil finds came the discovery of a major sour gas field at Waterton in 1957.

Alberta's economy boomed in the dozen years after Leduc. Many sectors developed or expanded vastly to support the mushrooming oil industry. The need for trained personnel was acute on rigs, refineries, pipelines and the many other facilities that sprang up overnight.

Many returning men from the war effort found work in the oil patch. Specialized training was also necessary and the Petroleum Industry Training Service grew up to meet this need. It held its first classes at the University of Alberta in 1949 in conjunction with the university, the Canadian Petroleum Association, the Canadian Association of Oilwell Drilling Contractors and the Energy Resources Conservation Board.

On January 2, 1950, petroleum displaced liquor as the single largest income generator for the Alberta government. By the end of that year, over 50 million dollars poured into the Alberta treasury from petroleum leases, reservation fees and production royalties.

The story of John A. Gow, a farmer near Barons, Alberta, puts the boom into perspective. On November 22, 1950, when asked if he had seen the gusher that was spewing oil on his land he replied, "Not yet. No point in getting excited. I've waited 36 years for this day, and I reckon the oil will keep for another hour or two until I've had my lunch."

Still, the boom was good news for the economy. Money was saved as well as made. By December 31, 1952, prairie consumers had paid $60 million less for their petroleum products by using local product than if they had used imported oil from foreign sources. Slowly, petroleum products invaded Canadian homes. Natural gas surpassed wood as the most popular fuel in homes in 1955, replaced hydro-electric power in 1960 and took over coal's place on the list in 1964.

Finally, in 1957, Alberta's petroleum production allowed Canada to claim technical self-sufficiency in oil and gas. A production potential of 900 000 barrels per day exceeded consumption of 750 000 barrels per day. In 1959, Alberta produced the billionth barrel of provincial oil. And in 1960, the Alberta government's income from the oil industry passed $1 billion.

Shell Oil Canada rig drilling at Jumping Pound, Alberta, c. 1952
(Harry Pollard/Provincial Archives of Alberta/P2945)

Oil workers and family at Leduc Calmar Co. well, west of Leduc, 1947
(Harry Pollard/Provincial Archives of Alberta/P1400)

On May 14, 1994, the people of the town of Turner Valley hosted a celebration honoring the 80th anniversary of the discovery of Alberta's first commercial oil and gas field. Among the old-timers was 98-year-old retired Turner Valley barber Lester Cox. He recalled the early days before oil, when "in 1914 you couldn't buy a job." Things were so tough that he moved to the oil patch to find customers for his scissors and razor. Oil came to the rescue. Right through the 1930s, the petroleum industry made a big difference. "Without oil, Alberta would be a lot like Saskatchewan."

Oil certainly did change Alberta. In the same way that Atlantic No. 3 blowout caused a sensation, so did the massive process of developing oil in the province. Yet no one took the time to look at the effects of oil pollution on the farmer's land. No one has to this date.

For the most part, the same holds true for an investigation into the impact of oil on Alberta. Economic fortunes were made and lost by individual investors, but the heady effect of the boom still lives on in Canada's richest province. Politics changed in a poor province once its economic clout allowed it to challenge the federal government in Ottawa and the powerful provincial legislators in Ontario and Quebec. Social conditions improved dramatically. Excellent education and health care became widely available.

But cashing in the inheritance had its costs. It took decades for Albertans to see the long-term consequences of a development process that put progress first. Half a century after Leduc, the bills for the boom still arrive daily. Pollution threatens quality of life even though Alberta is a leader in the environmental field. Although income from petroleum seemed vast during the booms, fear of provincial and federal debt during recessions prompted politicians to drastically cut into the rich programs that accompanied the wealth brought by the good times.

The fabulous success of the Leduc No. 1 well and the spectacular accident at Atlantic No. 3 are two related facets of Alberta's oil story. The bounty and peril that accompany petroleum are inextricably related and tie the story of this fascinating business and the Canadian people together in the past, present and future.

The Silent Highway

. . . transporting oil and gas

Train leaving Nisku with loaded tankers of Leduc oil, 1947
(Harry Pollard/Provincial Archives of Alberta/P1444)

Shock and horror coursed through American veins when they saw film footage of Japanese warplanes bombing Pearl Harbor. That December 7, 1941 attack on the warships in Hawaii pushed the United States into World War II. The threat was suddenly real. The Pacific coast was in immediate danger. The land of the Rising Sun was invading. The Red hordes poised, ready to conquer Alaska, then Canada and the lower 48 states. And so in 1942 Americans began building the most expensive oil pipeline in history across part of the Canadian north. With typical military jargon, they secretly code-named it the "Canol" project.

In late 1943, a special committee of the United States Senate met to investigate the National Defense Program. These men were upset at the cost and logic of building a pipeline that had not provided a drop of oil for the war effort. When Major General S. Clair Streett, Commanding General of the Second Air Force took the stand, the committee members pounced.

Senator Kilgore: "Now, if worst came to worst . . . what assistance would that 20 000 gallons be? . . . You couldn't do much fighting on 20 000 gallons."

General Streett: "It would help a lot."

Senator Kilgore: "It would help, but you couldn't maintain a scrap."

General Streett: "Not with large forces . . . . Gentlemen, do you realize what the position of the United States was at that time?"

Senator Kilgore: "Oh, yes."

General Streett: "It was one of the most precarious positions the United States

has ever been in . . . . The Japs had things pretty well their own way. Things were going very, very poorly; there wasn't a good thing to be said about the situation at all."

"It looked as though Alaska was next; in fact, we figured it would be, and if Alaska was next and the Japanese had control of the seas, we didn't see how we could move much of anything into Alaska, and if the Japs actually got on the Alaskan mainland, our Northwest was easily reached."

The grilling went on and on. Did anyone know how long it would take to build the pipeline? Who was counting the costs? What about the logistics? Why in the world did they spend so much money and human effort to build a pipeline?

Senator Ferguson: "How were going to get the pipe in? How were you going to get the refinery in?"

General Streett: "Sir, I didn't know, honestly. I was just hoping to God that somebody could do it, and if they could do it, it was velvet. That is my point. We were in pretty dire straits there. The War Department was the gloomiest place on the face of the earth, and naturally, being called in as somebody who knew the general situation out there, and asked, 'Could we use it?' – that was practically all I was asked – I said we could use it . . ."

Mr. Halley (interrupting): "I have been sitting here listening to the story, that we were losing the war, and I fail to follow the logic of that. I wish you would show the connection between the fact that we were losing the war and the fact that we ought to build $130 000 000 worth of oil wells and pipe line and refinery in Alaska . . . "

General Streett: "The question of cost was something I knew nothing about . . ."

Senator Kilgore: "In my opinion there was this danger: If your theory went through and the Japs did go in there, what development we did do would be for Japanese benefit, because you couldn't pull those wells up and you couldn't take that pipe line out, and we were probably putting gasoline in there for the Japanese."

General Streett: "That is possible."

And so the general admitted that they built a pipeline without counting the costs, how long it would take to complete, or if it might fall into enemy hands. The senators ignored the fact that the pipeline was completely in Canadian territory, from Norman Wells in the Northwest Territories to Whitehorse in the Yukon. As far as they were concerned, the United States owned the North during World War II.

The Canol pipeline was an incredible project, even for a wartime emergency. Almost 1 000 kilometres long, the 10-centimetre pipe lay on top of the ground because Norman Wells oil flowed even at very low temperatures. Imperial Oil drilled at least 40 additional wells at its Norman Wells field to supply the overnight demand. Across the mountains, the Whitehorse refinery sprang up with a capacity of 3 000 barrels per day. Equipment for it came from Ontario, Texas and California.

Finally, on February 16, 1944, a welder finished the last joint. By the time the first oil arrived in Whitehorse

in April, the Japanese threat was gone. The refinery operated for less than a year and closed in March, 1945. *Newsweek* magazine called it "the white elephant of Whitehorse."

The Canol project eventually cost $134 million and produced 976 764 barrels of oil. At $137.[19] American per barrel, Norman Wells oil supplied a very expensive source of gasoline for the Allied forces in World War II. In less than three years a pipeline went from an urgent necessity to a useless folly.

As strange as this project seems, it is merely one of the more unusual stories of the life of a pipeline. Although the petroleum industry is famous for its cyclical nature, its booms and busts, its fortunes and follies, no part of the process depends more on stability than the transportation system.

Petroleum is a curious resource. It seldom occurs in convenient locations. As a liquid or a gas, it is awkward to handle and store. It can be lethal to produce or use. It requires processing, scrubbing and breaking down into component parts. A multi-purpose product, it requires a complex system to use its incredible power.

Although both are valuable products, oil is heavy and gas is explosive. It is not economical to transport petroleum products long distances using conventional methods. Railways usually handle heavy and awkward materials but they cannot efficiently transport oil and gas across a continent. Railway tankers only move products regionally.

Ocean tankers are the most practical way to move oil. They provide an international link to cheap oil in the Middle East and South America and multinational companies have long relied on them as the backbone of their far-flung empires. Tankers go anywhere there is a port. Each shipment can go from a different oil field to a different market, depending on demand. But ships must operate on water.

The most economical way to move oil on land is by pipeline. Pipelines also have serious limitations. They require a consistent source and market and stable prices. Although relatively cheap to operate, they are expensive to build.

At first glance, pipelines seem mundane. They lie buried under the earth, connecting producers and consumers many thousands of kilometres apart – just silent carriers of precious goods. Over time, they have become expensive, politically charged and controversial. Most of all, they are an important underpinning for our society.

As history shows, pipelines are anything but boring. It all began innocently enough more than 100 years ago. Most people forget that Canada pioneered the business of getting gas and oil to market in pipes.

In 1853, a cast-iron pipeline began moving natural gas 25 kilometres to Trois-Rivières. As the longest line of its day, it provided fuel for street lights. Nine years later, the world's first oil pipeline began moving Petrolia crude to the Sarnia refinery in southwestern Ontario.

In Western Canada, Eugene Coste's Canadian Western Natural Gas, Light, Heat and Power Company Limited built a 275-kilometre long gas pipeline from Bow Island to Calgary. It began supplying the city in July, 1912. In 1921, Canadian Western began purchasing natural gas from Royalite's Turner Valley gas field to augment its rapidly diminishing supplies. Further north, a pipeline began moving Viking-Kinsella gas to Edmonton in 1923.

In 1924, the Royalite No. 4 oil well tapped into a new reservoir of gas in the Turner Valley field and spurred on yet another boom. As a result, Imperial Oil built Alberta's first oil pipeline from Turner Valley to Calgary in 1925. Then in 1936, the Turner Valley Royalties No. 1 well found crude oil below the wet gas and sparked the third Turner Valley boom. This important discovery proved the field's petroleum potential and produced more oil than the regional market could consume.

These early pipelines all held one thing in common: they supplied local markets. But the 1936 oil discovery in Turner Valley changed the politics of Canadian pipelines forever. Instead of being content with providing oil and gas to regional customers, producers began clamoring for big capacity pipelines to take their products from western Canada to the Great Lakes and beyond.

Overnight, pipelining became an international business with high stakes for all. Independent regional producers fought with giant multinational corporations for access to markets. As the previous chapter explains, World War II only temporarily solved this battle between conflicting oil interests. For the duration of the war, petro-

Hauling pipe in the Turn Valley oil field, c. 1929 (Glenbow Archives/NA-711-5)

Workmen installing threaded pipeline by hand from Turner Valley to Calgary, 1920s
(Provincial Archives of Alberta/PAA-85.248 Box 39)

Workers installing North Western Utilities pipeline in Edmonton which supplied Viking-Kinsella gas to the Alberta Capital, August 21, 1923 (McDermid Studios/Glenbow Archives/ND-3-2063)

leum was in high demand everywhere and pipeline projects took a back seat to immediate security concerns.

The one exception was the expensive pipeline from Norman Wells to Whitehorse. A creative wartime engineering success, it was fabulously expensive but turned out to be less than critical to the war effort. Decades later the code name "Canol" still lies shrouded in mystery. Most reporters routinely parroted the American army propaganda that "Canol" was just a contraction of the words "Canada" and "oil." But the project was not called Can–oil.

Politics played a crucial part in

this clandestine military operation in the Canadian far North. With Ottawa's blessing, Americans invaded the Northwest, building the Alaska Highway, strategically placed airstrips and the pipeline leading from nowhere and going to nowhere.

And that led to the cover-up. The American military did not want Canadians to know that the North was under virtual siege by their southern neighbors, so they propagated the myth of Canol.

In fact, Canol was an acronym for Canadian American Northern Oil Lines. Years later, Ted Link Jr. heard it explained on the golf course. Patricia

Before railway tankers and pipeline, wooden barrels were the only way to move oil.
Wainwright, Alberta, 1925 (Provincial Archives of Alberta/A10803)

Barry, the author of a major study on Canol, got the same definition from an old-timer in the North. The most credible witness to date is probably an octogenarian named Bernard (Barney) Middleton.

Barney's job was to make things happen in Norman Wells. He wrote dozens of recurring reports, explaining in detail every aspect of the Canol project. In fact, it was hard to get real work done. Emil Schudel, comptroller for the project, once quipped: "If the war goes on much longer we're going to have to quit because we're running out of paper."

As right-hand man to Ted Link Sr. on Imperial Oil's part of the Canol project, Barney was intimately involved with every part of the operation. Even though he never saw Canol's definition written on paper, he is adamant about the mysterious acronym.

According to him, secrecy was the order of the day: "Everybody from the FBI to the CIA was involved up there." And so the Americans obfuscated the meaning of Canol with great success.

Less successful was the war effort to find oil elsewhere in Canada. As the result of intensive geological, geophysical and drilling efforts in the late 1940s, massive oil fields finally came on stream all at once.

But as of 1947, there were few pipelines in Canada. Some moved oil and gas from Turner Valley to Calgary. A crude oil line transported offshore oil from Portland, Maine to Montreal and some short lines brought oil from the United States to central Canadian refineries. These various lines totalled less than 700 kilometres in length.

After 1947, pipeline construction became a booming industry. By

Gas flared into a coulee at Turner Valley town prompted locals to call the site "Hell's Half Acre."
Note frost on the pipe. 1926 (Glenbow Archives/NA-1716-5)

An early pipeline ditching machine digging south of Calgary in the 1930s
(Harry Pollard/Provincial Archives of Alberta/P1974)

Truck that transported oil and the train car that replaced it, 1947
(Harry Pollard/Provincial Archives of Alberta/PA415/3)

1960, the backbone of the Canadian skeleton of oil and gas lines was in place. Within another decade, tens of thousands of kilometres of line snaked around Canada, silently delivering oil and gas to millions of customers.

Pipelines were soon routine operations. At the end of 1949, with new markets ready to reach out to Alberta oil and gas, Carl Nickle of the *Daily Oil Bulletin* wrote in his annual review:

In the mounting worldwide war between ideologies, western Canadian oil and gas will increase its importance in the security planning of the democracies.

1949 brought self-sufficiency in petroleum to the Prairie Provinces, and the beginning of "growing pains" for the fast expanding oil industry.

Anything as monumental as continental pipelines required regulation. In 1949, the Canadian Parliament passed the *Pipelines Act of Canada* to control the new transportation system. Right on the heels of the regulatory body, Imperial Oil and the Interprovincial Pipe Line began drafting plans for a pipe from Alberta to the Great Lakes.

During 1950, work crews carved their way across the prairies and between the spring and the fall installed the oil line from Edmonton, Alberta, to Superior, Wisconsin, in just 150 days. Finally, on October 14, oil from Redwater began moving across the continent to Wisconsin through the 1 800-kilometre line.

This project marked the beginning of the export era. As a result, Canada became involved in the intern-ational oil industry. Instead of just producing for its own consumption, Canada began serving continental markets.

The remainder of the pipeline story could be simply a list of impressive engineering projects. Some lines posed unprecedented technical problems. Pipeline crews successfully installed lines through mountains, under rivers and lakes, and through parts of the most inhospitable territory that makes up this vast country. Within a decade, pipelines to anywhere seemed possible.

But more complex issues lay around the corner. Although engineers and promoters sometimes describe pipelines as engineering accomplishments, many other factors often come into play. Carl Nickle lobbied hard in his December 31, 1952 year-end review, for a rational approach to petroleum development.

Will the matching of gas reserves to markets be based upon sound economics and common sense, designed to achieve maximum returns to the producer and his provincial governments, and the minimum possible fuel costs to Canadian consumers, or will the mating be based upon emotionalism and nationalism without regard to economics? Will Canada and the United States deal with gas, as they largely are with crude oil, on the basis of continental unitization for the benefit of both producers and consumers of the neighbor nations?

In the answers to those questions rests a big part of the future of Western Canada's oil and gas industry,

Loading a tanker at Nisku with Leduc oil, 1950
(Harry Pollard/Provincial Archives of Alberta/P1540)

and the role that industry will continue to play in the economic well-being of the West, the nation and the continent . . . . The year ahead, and the years beyond, offer a vaster challenge, and vaster opportunity, to the trio of the public, government and the oil industry."

Once petroleum marketing became a continental project, the complications increased exponentially. For example, although Canadian governments quickly approved oil exports to the United States, they were hesitant to throw open the valves and allow natural gas to flow south. Calls for a rational approach to this subject fell on deaf ears. Canadians did not consider gas in the same light as oil. Perhaps gas provokes more emotional response because it heats homes during long, cold winters. Regardless of the reason, legislators reacted differently.

Finally, in 1949, Alberta passed the *Gas Resources Preservation Act* which, after considering the province's needs 30 years into the future, approved exports.

Even that gas reserve was not enough to clear the way for the first natural gas pipeline to central Canada. In the 1870s, Ottawa demanded that the Canadian Pacific Railway lay its

Welders joining pipe on IPL line near Regina, Saskatchewan, August 5, 1950
(Harry Pollard/Provincial Archives of Alberta/P1366)

Welders joining pipe on IPL line near Maryfield, Sask., August 8, 1950
(Harry Pollard/Provincial Archives of Alberta/P1414)

track around the north shore of the Great Lakes, all on Canadian soil. The same sentiments arose when promoters attempted to build the gas line through the United States to central Canada.

As a result, TransCanada Pipe Lines Limited promised in its charter "that the main pipeline or lines, either for transmission or transportation of gas or oil, shall be located entirely within Canada." All seemed ready for construction in the early 1950s.

Nothing is that simple in Canadian history. First Alberta had to approve the exports and the regulators in the United States balked at the gas reserves and the financing. Ottawa eventually had to create a crown corporation to build the financially risky 1 100-kilometre portion of the pipeline across the top of the Great Lakes and then lease it back to Trans-Canada Pipe Lines.

In the end, the 3 500-kilometre long line was controversial. Debate over the pipeline project on the floor of the House of Commons dragged on until the Liberals used closure to force a vote. Labelled as the "Great Pipeline

Ditcher preparing trench near Maryfield, Sask., August 8, 1950
(Harry Pollard/Provincial Archives of Alberta/P1417)

Debate" by the press, it contributed to the fall of the Liberal government in the 1957 election. On October 27, 1958, gas from the prairies began moving under its own pressure from Alberta through the world's longest natural gas pipeline.

Not all projects were that controversial. Because of the increasing demand for permission to export oil and gas, Ottawa commissioned a study into Canada's petroleum reserves. In 1958, the Borden Royal Commission on Energy recommended that Ottawa create

the National Energy Board (NEB). That it did in 1959 and the Board still oversees the development of Canadian energy projects in the interests of the Canadian public.

The 1950s and 1960s witnessed the construction of many other pipelines in Canada. In 1953, the Trans-Mountain oil pipeline started moving oil from Edmonton to Washington State and Vancouver. In 1957, Westcoast Transmission built a gas pipeline from northeastern British Columbia to Vancouver.

Lowering line into trench near Maryfield, Sask., Aug. 8, 1950
(Harry Pollard/Provincial Archives of Alberta/P1420)

Official opening of the first Canadian long distance pipeline near Edmonton, Alberta, October 4, 1950
(Harry Pollard/Provincial Archives of Alberta/P3407

Crowd of spectators at official opening of the first Canadian long distance pipeline near Edmonton, Alberta, October 4, 1950
(Harry Pollard/Provincial Archives of Alberta/P3409)

During the 1950s, Alberta Gas Trunk Line (now called NOVA Corporation) sprang up to gather Alberta natural gas and move it to the provincial borders for export. In 1961, Alberta and Southern began exporting natural gas through a large-diameter pipeline from Alberta to San Francisco.

During the 1960s Alberta production increased greatly. In 1967, the Great Canadian Oil Sands (now called Suncor) began producing synthetic crude from tar sand at its Fort McMurray refinery. That same year, the Alberta oil and gas economy boomed due to problems in the Middle East and the threat to the North American supplies.

In the period from 1960 to 1969, Canadian gas demand more than doubled. And the petroleum industry stayed ahead of the demand. Finally, on August 31, 1970, Alberta celebrated its first ever "Million Barrels Daily" milestone.

The year 1970 marked another important turning point for the Canadian oil industry. With export questions resolved and regulatory bodies in place to control the development process, the oil patch seemed ready to roll into the future without a care in the world.

The relatively peaceful era of petroleum development in Canada came to an abrupt end in the early 1970s. When the international price of oil began climbing as the result of Middle East wars and the actions of the Organization of Petroleum Exporting Countries (OPEC), the fight for profits became intense.

Central Canada looked to the prairie producers for relief from high oil prices just as the poor-producing provinces saw the clouds open and rain down profits from higher oil prices. Western premiers began flexing their political and economic muscles in 1974 and the battle was on. Pipelines linked central Canada and the West in a complex supply-and-demand relationship that remains testy to this day.

During the early 1970s, another debate sprang up that prevented prompt construction of another long pipeline project. The North, long a proven oil producer, held out the promise of even more oil for the rest of the continent after the discovery of oil at Prudhoe Bay in Alaska in 1968.

Early feasibility studies recommended that oil and gas pipelines run from the American and Canadian Arctic down the Mackenzie Valley and through Alberta and into the United States. Various companies competed for the right to build the lines. Finally, in 1974, Ottawa appointed the Mackenzie Valley Pipeline Inquiry to investigate the complex development process. Many issues required scrutiny.

In an effort to develop the first new American oil field in many decades, Congress narrowly approved the Trans-Alaskan Pipeline System in 1972. With OPEC flexing its economic muscle, the United States was anxious to decrease its dependence on foreign supplies. Environmental concerns and Native land claims negotiations stalled the project for years.

The inquiry in the Canadian North finally released its report in 1977. Commission chairman Justice Thomas

Berger called for a moratorium on development in the North. He recommended against a pipeline across the Northern Yukon due to its environmental damage, and that the Mackenzie Valley line be postponed for 10 years. The moratorium would allow for more exploration, for settling land claims and for creating systems to ensure that Natives could adapt to the effect of the changes on their lives.

Finally, almost a decade later, on May 17, 1985, the Norman Wells Oil Field Expansion and Pipeline were officially opened. With a capacity of 25 000 barrels per day, it now moves crude oil from Norman Wells to Zama, Alberta.

The Mackenzie Valley Pipeline Inquiry was not the first project to undergo close environmental scrutiny. One of the first projects to come under harsh criticism was the early gas pipeline from Alberta to California. In that case, the concern was not over the environmental effects of the pipeline, but with the smell of sour gas that greeted the customer. Although the legal limit of sulphur was strictly adhered to, California consumers forced the Canadian producers to lower their maximum sulphur content still further before they would buy the product.

As the industry has learned, the realm of environmental affairs includes many surprises that an engineering study does not necessarily predict. The Alberta government changed too and in 1971 it broadened the Energy Resources Conservation Board's mandate to include the responsibility "to control pollution and ensure environmental conservation in the exploration for, processing, development and trans-

portation of energy resources and energy."

But much damage had already been done. As farmer Peter Lewington documents in his book, the early pipeline companies wreaked havoc on their right-of-way when they installed lines in the early days. And for years, the land owners got nothing but a deaf ear from the companies and the government agencies appointed to protect the public interest.

Unfortunately, it took decades to resolve some of these problems. For example, on Christmas Eve, 1976, a lawyer for a pipeline summarized his company's perspective on the case before the court with a harshly legalistic conclusion.

"I concede that Lewington and O'Neil are doing this out of a deeply held conviction, but they are turning it into a cause célèbre. Right or wrong, a pipeline company can go into a property and turn it into a wasteland."

Eventually these farmers did receive a satisfactory resolution to the problems that plagued them for decades. Lewington now praises the companies for the changes in their operational practices.

"Some energy companies have clearly demonstrated that they can prosper without injuriously affecting a landowner or the land. They have grasped the essential truth — and the economic benefits — of good planning and practices that mitigate damage. It is doubly rewarding for them to find that doing what is morally and

aesthetically desirable is also the most profitable course of action. And that has to be a very important benefit of bringing democracy to the oil patch."

Canadian pipeline companies are now among the most environmentally responsible in the world. They spend millions of dollars annually preventing and repairing the damage they do to the land they pass through. The new breed of men and women not only take care of the land because they must; these environmental managers for petroleum companies are often university-trained biologists and other natural scientists with a genuine respect for the land, air and water. In addition to all their other complexities, the many thousands of kilometres of Invisible oil and gas pipelines are also potentially dangerous. Although considerably safer than other methods of

transporting petroleum, this industry has its share of accidents.

In general, the oil patch is a potentially deadly place to work. In the early years only coal mining was classified as more accident prone than the oil and gas sector. At first, the men who worked on oil wells and in the production, refining and distribution system were not aware of the hazards. For example, highly-pressured gas caused the death of two men at a drilling rig in Turner Valley on March 31, 1929.

The casing rose out of the well and toppled on the workers as they ran for safety. A separator explosion killed another man that year. Others died when crushed by timbers and when rigs collapsed on them. Falling tools in a rotary rig killed another man when they hit him on the head. In the days before hard hats, safety was lax and

Gas well explosion at Freehold No. 1 well, December, 1928
(Glenbow Archives/NA-711-95)

accidents more common.

As time passed, informal and company-sponsored training programs taught the oil rig workers the rudiments of safety standards. But sour gas persisted as the greatest threat to life. Even with experience the men did not fully understand the danger of hydrogen sulphide laced gas. The fact that they always smelled the rotten egg smell sometimes made them feel invincible to its effects.

The army surplus gas masks the companies provided were useless so the men invented their own ways to stay alive. Some consciously leaned away from the valve they were opening, hoping they would fall out of danger's way in case the gas knocked them out. Some learned from the deaths of their friends and tied a rope around their waists so others could pull them to safety when they fell unconscious.

But only better planning could have saved the lives of two men at the Turner Valley gas plant in 1943. When the fans quit one man went into the building to reset the switches and fell prey to the deadly gas. The man who went to save him was knocked out too. Only after this accident did the company move the switches outside, where they should have been all along.

Gas processing plants like the early one in Turner Valley scrubbed the deadly hydrogen sulphide from the natural gas. Pipelines then transported it to consumers. But even these buried transmission systems were not without their spectacular accidents.

One such incident happened on the TransCanada pipeline on Christmas Eve of 1957 near Dryden, Ontario. According to William Kilbourn, the explosion created "a tremendous roar, the ditch heaved, the cover of soil and rock flew upwards, and the pilot of a Trans Canada Airlines [now called Air Canada] aircraft saw a long flash of light leap from the earth below him. The longest pipeline break in history, some three and a half miles, took place instantly."

That pipeline met all prevailing standards. As a result of that incident the National Energy Board began working with the Canadian Standards Association to create pipelines better suited to the harsh Canadian environment. Today's pipeline systems are even more sophisticated than ever before, operating with an enviable safety and reliability record.

In retrospect, the story of the silent highways that transport oil and gas from producers to consumers is complex and sometimes full of intrigue and danger. Within the space of four decades, our society has inextricably linked itself through a complicated web of gathering and distribution pipelines. These lines now move valuable petroleum products around the continent through an intricate system most North Americans take for granted. As the arteries of the energy life support system, pipelines are pivotal to our economy.

Hortonspheres at Royalite's Turner Valley gas plant, Sept. 9, 1952
(Harry Pollard/Provincial Archives of Alberta/P2604)

The Earth Movers

. . . digging deep into the oil sands

Founder Karl Clark worked out how to separate oil from the sands in the 1920s, in a laboratory at the University of Alberta. (Syncrude Canada Ltd.)

On a crisp late-September day, 400 kilometres northeast of Edmonton in Fort McMurray, the Progressive Conservatives who ran the federal and provincial governments did their best to make the Christmas of 1988 come early. A $100-million grant for preliminary engineering headed a $2.8-billion list of promises from delegations led by the province's premier, Don Getty, and Joe Clark (who took time off from his role as Canadian foreign affairs minister to fill in at the gala event for Prime Minister Brian Mulroney).

Broadcasting live via space-satellite links set up at a hilltop motel above the oil sands city, the governments announced they would back OSLO and the giant Syncrude project. Grants, loan guarantees, interest-cost supports, part-ownership and oil-price subsidies called "indexed development incentives" would cover 60 percent of the $4-billion cost forecast by OSLO.

Lapel pins sporting "OSLO" circulated, which stood for "other six lease owners." The name referred to 200 square kilometres of oil-rich sand near enough to the surface for strip-mining. Ten years earlier, the blue-chip corporate owners had put into production the giant which sired the word megaproject: Syncrude Canada Ltd. OSLO – led by Imperial Oil Ltd.'s supply development arm, Esso Resources, and backed by Canadian Occidental Petroleum Ltd., Gulf Canada Resources Ltd., Petro-Canada and PanCanadian Petroleum – spelled "big."

Over OSLO's 35-year lifespan, book-length press kits promised that building and running the complex up to its 77 000-barrels-a-day capacity would create 6 000 construction jobs, yield permanent employment for 2 770 plant operators and eventually pay $13 billion in wages and salaries. Getty, usually soft-spoken and re-

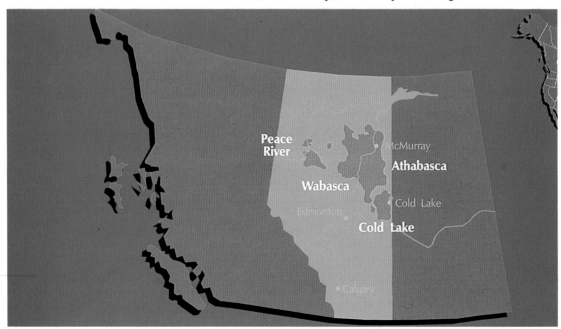

Resource Carpet: Vast beds of oil mixed with sand, first noticed in the 18th century, inspired generations of efforts to exploit energy awaiting the right technologies in astronomical quantities (Imperial Oil Ltd.)

New Generation: Retrained operators now run Suncor's plant with keyboards at a handful of video display terminals, employing the latest in digital computer technology (Suncor Inc.)

Tumblers: Conditioning drums churn the oil sands while adding hot water and steam, in the giant commercial-scale application of Karl Clark's process. (Suncor Inc.)

nowned for disliking political bally-hoo, grew eloquent as the promises multiplied.

"The noose of foreign-dominated oil supplies will never tighten around Canada's neck," the premier declared in describing OSLO's benefits for Canadians who would not climb directly on to its payroll. "OSLO will be successful."

The occasion carried away even the normally reserved Esso oil sands manager who doubled as OSLO's chairman. Gordon Willmon, who came into the announcement ceremony saying the complex would be built if the oil price became right, finished by saying he believed it would be. He declared the governments gave so much help that "this is a viable, economic project under any realistic pricing outlook."

Down the hill on the same day, the view looked less rosy to wage-earners, homemakers and shopkeepers who congregated at Fort McMurray's Heritage Park for a picnic to celebrate the 10th anniversary of Syncrude starting production. Amid relics of the fur trade that founded the outpost on the Athabasca River in the 18th century – and machinery left by oil sands experiments that also spanned a good part of the site's history – the community had no trouble keeping politicians in perspective.

Hilda Hinchcliff, whose son-in-law had been out of work for two years since the 1986 oil-price collapse, said "we sit tight and just wait to see what happens."

Another mother, Mary Samson, declared "this is another political scene" and forecasts of big projects are "pretty run of the mill around here." She predicted – correctly – that a federal election could not be far off.

Oil sands plant worker Garry Hennings said, "I believe they're going to go ahead with the engineering, as the basic work to find out if they've got something in tune with the times." Like most of his neighbors, Hennings paid attention to the cautious Willmon at the beginning of the OSLO news conference, rather than the enthusiast who emerged at the climax. The blue-collar worker agreed with Willmon's caution.

"You can't just jump in and say we're going to build a plant. I'm not going to go out and buy tons of property on speculation that a housing boom for new workers is just around the corner."

A Syncrude crew supervisor, Buzz Erlinger-Ford, observed "there's not many people who have illusions about getting rich quick in Fort McMurray. Anybody that's been around here a while is going to be cautious."

Within six months, the Conservatives won reelection and Imperial put the moment of euphoria behind it and confirmed that the rank and file in Fort McMurray had it right. In its annual report for 1988, Canada's biggest oil company classed OSLO with Arctic and other costly, big and remote projects as "long-term opportunities." Why? Because "there is still a sizable surplus of oil-production capacity in the world, and, as long as that situation exists, the market will remain soft. For the short term, therefore, Imperial will manage its business on the premise that oil prices will remain weak and volatile over the next several years." Imperial drew no quarrels from its corporate peers.

A year later, in early 1990, the federal government cancelled the next, $140-million step in aid to OSLO. Petro-Canada chairman Bill Hopper called the project "a dog." He quickly apologized but kicked in no more money. The rest of the partners more quietly shelved OSLO when the $100 million in preliminary engineering turned in a construction estimate of $4.5 billion, or a 12.5 percent overrun that let them drop out under their agreement with the governments.

An industry survey by the National Energy Board reported a consensus that soft world oil markets had pushed the earliest conceivable date to start construction back to 1999. OSLO's design team quietly disbanded, ending more than 100 jobs.

The workers and their dependants in Fort McMurray intended their employers no slight when they refused to believe in OSLO from the start. No offense was taken. The community knew its history. It paid the industry a bigger compliment than the politicians, by recognizing the scale of technical and economic accomplishment required to build an oil sands production complex.

Big visions come naturally to the oil sands – and so do prolonged struggles trying to put them into practice. The resource carpets 62 500 square kilometres of northern Alberta, or an area 100 times the size of even the Canadian petroleum industry's home-base, Calgary. Although only 10 percent of the deposits lie close enough to the surface for mining, even that minority represents a 250-billion-barrel fortune of Middle Eastern proportions.

But the oil is heavy, requires extensive processing into a state ready for refining and then still has far to go to markets. Oil sands entrepreneurs also take on an environment that prepared the Canadian industry for its 1990s expansion into Siberia. While Fort McMurray lies far south of the Arctic Circle, the surroundings resemble the scrub forest around Inuvik just inside the northern treeline.

The oil sands occupy a region of harsh continental climate, far beyond the reach of the warm Chinook winds that flow out of the Rocky Mountains to moderate the climate of western Alberta and northeastern British Columbia. Even after nearly 20 years as home to large-scale industrial development, Fort McMurray startles newcomers with its remoteness and with its months-long cold snaps near -40°C. On leaving town headed north, the first road sign announces to travellers that the next gas, food and lodging will be found in Fort Chipewyan, 280 kilometres of gravel and ice surfaces away.

Realization that the oil sands resisted get-rich-quick schemes dawned as soon as serious work began on tapping the resource. The blend of sand and tarry bitumen started inspiring imaginations almost as soon as explorers encountered it, in use as waterproofing for Native canoes, and seeping from exposed banks of the Athabasca river on hot days of long summer sunlight. OSLO followed a decades-old pattern of discovery, optimism and a return to realism after trials of the latest new

Separation cells yield oceans of bitumen from a "slurry" of oil sands and water. (Suncor Inc.)

Black Gold: Bitumen taken from the oil sands without processing into synthetic crude has the consistency of molasses but efficient production makes it pay. (Imperial Oil Ltd.)

Large, Long Experiments: Edmonton-area pilot plant began production tests in 1925. (Provincial Archives A3525)

ideas. The architect of oil sands technology, Karl Clark, pioneered this pattern. He also earned the right to have schools and streets named after him in Fort McMurray and Edmonton by inventing the process that eventually made production possible.

Clark joined an enterprise that already had a long pedigree when he moved to Edmonton in 1920 from his native Ontario. There he had worked as an engineer in Ottawa on ideas for using bitumen from northern Alberta to pave roads across Canada. The oil sands were entered into the nation's official resource inventory by 1875,

Promotion: Road paving raised the profile of the oil sands as a resource that could be used. (Provincial Archives A3399)

through records of the Geological Survey of Canada.

Ambitious promoters tried drilling the first well into the oil sands in 1894, at Athabasca Landing on the southern end of the deposits, about 160 kilometres north of Edmonton. When Alberta became a province in 1905, leaders such as the University of Alberta's founder, Henry Marshall Tory, made requests to the federal government for matching economic status through a transfer of resource-ownership rights. The oil sands played a prominent role in these requests.

Less than two years after Tory recruited Clark to probe the oil sands as the first paid employee of the U of A-based Alberta Research Council, the scientist sent Tory a gratifying letter. (His work and early oil sands history are preserved in correspondence collected by daughter Mary Clark Sheppard as *Oil Sands Scientist: The Letters of Karl A. Clark 1920–1949*, Edmonton, University of Alberta Press, 1989. Some excerpts from this book follow.) In the fall of 1921, Clark wrote, *Most of the purely inventive work has now been done.* Employing hot water, he had figured out a way to separate the oil from the sand cleanly. He wrote Tory that the work had progressed far enough to stimulate some business excitement:

Jumbo: Scoop shovels, increasingly in use as more efficient than bucketwheels, in turn dwarf the oversized trucks. (Suncor Inc.)

Back to the Future: In rediscovery of truck n' shovel mining techniques, loads of up to 240 tons are tipped into a crusher. (Suncor Inc.)

Big Bite: Crusher chews oil sands into convient morsels for further steps in the separation process.
(Suncor Inc.)

Just in Case: Suncor's layout includes a quiet pond filled with several Olympic swimming pools–worth of clean water for fighting fires, a major hazard of oil sands production with heat processes.
(Suncor Inc.)

A process which extracts bitumen and an experimental piece of road in which the tar sand bitumen was used would probably be sufficient to cause activity among promoters and stock peddlers.

Clark's cue was taken. But it was delivered chiefly by S.C. Ells, a former colleague who stayed in Ottawa to work on the oil sands for the federal government and became a political and professional rival. High-profile experiments were conducted, using the bitumen to pave streets in Edmonton and Jasper, then as far east as Ottawa and Petrolia, Ont. Clark's job rapidly came to include visits and correspondence with business promoters.

But he also warned Tory it would be a long way from theory to practice in northern Alberta. Clark wrote:

The real business concern, on the other hand, is that who would undertake the exploitation of tar sands will not be seriously interested until quite complete and satisfactory information bearing on all sides of the undertaking are available. Reputable financiers would not commit the capital.

Two years later, in 1923, the engineer bluntly posed the challenges that all oil sands projects eventually had to overcome, in a letter to a Toronto firm that expressed an interest in the increasingly famous resource. Clark observed:

There are four sub-questions which must be answered before it will be possible to give an answer to the main question of whether the bituminous sand deposits of northern Alberta can be profitably developed. What are the most advantageous locations in the bituminous sand area for undertaking commercial development?

How can the bituminous sand be mined or excavated cheaply, and at what price? What products can be refined from the crude bitumen content of the sands, what value have they, and what market can be expected for them? Information for answering these questions is not available.

For half a century, with the scientific community refining technology and seers like Tory urging on the process, the federal and provincial governments took on field demonstrations to keep industry interested in the oil sands. The federal government bought and operated a plant called Abasand, after early efforts by Denver promoters Max Ball and B.O. Jones ended in a fire in 1941. Fire burned down the plant again in 1945, ending a federal effort that irritated Clark and his fellow Albertans by rejecting Clark's hot-water refining technology.

The Alberta government laid out $725 000 – big money for the time but only one-third of Abasand's cost – to build its own demonstration plant at a site on the Athabasca River called "Bitumount." Within three years of starting production in 1948, the Alberta government commissioned Sidney Blair, an old friend of Clark who moved on to become an industry consultant in Toronto, to write a report declaring that Bitumount proved the oil sands' time had come.

Privately, Clark admitted that he had discovered, the hard way, other obstacles the industry needed to lick to make the old dream come true – starting with the climate and location. After decades of exploring the northern bush, the conditions came so naturally to him that he rarely mentioned them. But he saw them anew when he had to give some advice to an engineer recruited from Oklahoma to work on Bitumount:

You may be sort of supposing that roses bloom at Bitumount at Christmas just as in Tulsa. The open (ice-free) season on the Athabasca river is short . . . a maximum of six months for doing things using river transportation. At a place like Bitumount, men get very restless in October. The weather turns cold and hard frosts occur.

The discovery of abundant conventional oil in 1947 – in far easier conditions within convenient motoring distance of Edmonton at Leduc – made oil sands life seem even harder by contrast. In his long contacts with it, Clark discovered there was more to industry than dollars and technology. Business psychology and corporate culture mattered too. He told Blair:

Regarding the attitude of the oil companies to the tar sands, I told someone recently it was like that of kids to something new in the way of food; they don't like it before they have tasted it. If the tar and sand oil would only come up through a well, every-thing would be natural and they would be interested. But having to dig it up is just too much to contemplate. They are all interested and are scared there may be something in it, and do not want to be left out if that is so, but hope they will not be called upon to do something they are not used to.

When Clark died in 1966 at age 78 after a lifetime of research, Great Canadian Oil Sands and its principal owner, Suncor Inc., remained nine months away from starting up the first commercial production plant.

What finally made the industry bite? Two retired veterans of the international Sun organization's Canadian arm, Ned Gilbert and Bill Tisdall, recall Philadelphia headquarters nibbled in the 1940s but hesitated until it held an exercise which eventually became a fixture of oil sands planning.

Sun drew two lines on a chart. One line projected oil consumption in the United States. The second line projected supplies from American conventional wells. Where the two intersected showed the onset of shortages and a requirement for imports.

This is what gave sources like the oil sands a chance. It helped that Sun was owned by a family, the Pews of Philadelphia, who had a penchant for long-range thinking; an enduring hunger to secure enough oil for all of its own refining and retail operations; and the means to keep "far-out thinkers" working towards this self-sufficiency, Tisdall says.

Gilbert owed his career to the quest. Soon after hiring him in his na-

Era of Giants: Great Canadian Oil Sands inaugurated jumbo development that became known as the "megaproject."
(Suncor Inc.)

tive Wisconsin as a fledgling geologist, Sun sent him to Canada in 1944 as the Second World War lit fires under the hunt for supplies. He started in Nova Scotia with four professional peers near the end of their careers. He moved to Alberta in time for the postwar drilling flurry that found Leduc in 1947. For two years, Sun's western Canadian outpost – and Gilbert's home – was a room in Calgary's Palliser Hotel. There, "the best recommendation I ever made to Sun was to get into the oil sands."

By coincidence, an organization that later became Shell Canada – Canadian Oil Companies, also known as Whiterose – both shared pipeline interests with Sun and held oil sands property that supplied the Abasand plant. Sun bought control of a major lease in 1954 and made a deal to supply a plant proposed by Great Canadian Oil Sands (GCOS) in 1958. When GCOS won a long contest for an Alberta permit to build its plant in 1962 and needed money to proceed, Sun chairman J. Howard Pew had the family firm buy 81 percent control and launch construction.

As the work proceeded, the Sun-GCOS project pioneered another trait of oil sands projects – big cost overruns. Originally estimated at $110 million, the construction tab nearly tripled to $300 million by the time about 3 000 workers finished the job in 1967. Among others helped by the project, Calgary's Southern family had their fortunes boosted by an order for 600 of their work-camp trailers. (The temporary mini-city went far towards putting the Atco brand of portable housing on the map in international industry.)

Recruiting and housing small armies of personnel only scratches the surface of the tall orders involved in erecting an oil sands complex. The developers literally have to make the earth move, even before mining can begin.

Although considered close enough to the surface to rate as a prime target, the deposit in the 26 square kilometre Suncor site lies under up to six metres of northern muskeg or swamp, then more layers, as deep as 45 metres, of gravel, boulder, clay and silt "overburden." Before digging can begin into the layers of oil-bearing sands (35 to 65 metres thick), the overburden has to be stockpiled carefully enough to be put back to cover exhausted sections of the mine, in compliance with environmental reclamation laws.

An oil sands complex is also a factory. Turning the tarry ore into a product a refinery can use takes a nine-step process, employing techniques devised by Clark, then refined by generations of engineers. Conditioning drums add hot water and steam. Rock and clay is screened out. The concentrated "slurry" of oil sand and water goes to separation cells that skim off bitumen. Then a flotation process dilutes the tar with naphtha and puts the mixture into centrifuges that spin off impurities. Cookers heat batches to 500°C.

The hot material goes into "cokers." Poured into vessels and stockpiles, they yield vapors, solid coke, trace metals and sulphur. "Fractionaters" separate the vapors into naphtha, kerosene, gas-oil and the "light ends" such as propane and butane. The light ends and coke become fuel for the plant's heat processes. "Unifiers" add hydrogen to the remaining material, to remove sulphur and nitrogen and then reassemble the purified naphtha, kerosene and gas-oil into synthetic crude oil.

From the beginning, the secret of success in the oil sands has been to achieve economies of scale – to make big volumes of synthetic crude so that costs of the process can be spread as thinly as possible. At the Suncor site, where the deposits are 11 percent to 12 percent bitumen, two tons have to be dug up and processed to make each *barrel* of oil. The process requires a gargantuan size for virtually every piece of equipment.

Mine trucks, with price tags well into the seven-figure range and which carry up to 240-ton loads arrive in pieces; they take tractor-trailer units to transport them to the site. The excavating equipment – 10-storey-high bucketwheels or scoops on rings that eat into the oil sands the way a circular saw bites through wood, and even taller draglines that swing mammoth shovels on wire ropes – in turn dwarfs the trucks.

In the processing plant, the pipes, drums, vessels, valves, towers, stacks and control systems attain a scale to match the mine. Finally, to collect and purify the wastes, lake-sized cooling and settling ponds were developed by piling up 100-metre dikes of the biggest by-product, the sand, which is scrubbed virtually antiseptically clean by the plant processes.

For Suncor's pioneer commercial oil sands plant, construction only began the work. Operating the complex became a case of learning on the job – and often the hard way, recalls Tisdall, an engineer who helped piece together the mine. Keeping even the smaller pieces going posed challenges.

Special alloys had to be developed, for example, to harden the excavating scoops because the abrasive oil sand rapidly wore down ordinary tool

Original Equipment: Bucketwheel excavator rapidly dug out oil sands, moving ever farther into the deposits on its own all-terrain steel treads. (Suncor Inc.)

Big Foot: Tread plates for a dragline make big impressions.
(Syncrude Canada Ltd.)

steel. Blasting with explosives was introduced to "fluff" the deposits for easier digging in summer and to slow natural freezing in the long, cold winters.

Changing a gigantic truck tire also turned to be an endeavor of heroic proportions, requiring a forklift to take off the wheel and heated garages to soften up the synthetic rubber enough to work with it. Ordinary electrical insulation cracked in -40°C temperatures, causing crippling shorts in power and control systems. Tractors had to be warmed up before performing routine maintenance.

In the glare of the northern sun off the wide vistas of treeless snow in the mine "you had to get good sunglasses for the bucket wheel operators," recalls Tisdall. "And the purchasing department always wanted to know *why*. These were things the bean counters hadn't figured on. Costs they hadn't anticipated went

on and on. You saved a dime a yard (of excavation) here, a nickel there. Little by little, you made it work."

After two decades of experience, with oil prices gyrating while expenses stayed at inflated levels (just one of the jumbo truck tires cost $10 000), in 1985 Suncor vice-president Bill Oliver called the oil sands "as big a risk as ever." The gamble has become so great that no one company would ever take it alone again, he told a press tour.

"No one wants to put a lot of eggs all in one basket," agreed vice-president Mike Supple. Nine years later, Suncor again underlined the giant scale of commitment it takes to develop an oil sands mine. The company announced a $175 million, three-year program of adding technology to cut emissions and improve efficiency at its team and electric power facilities.

No other company ever has tried to duplicate Suncor's feat of tack-

One Step at a Time: With the processing plant in the distance, a dragline starts the oil sands on their way to synthetic crude by heaping the deposits into "windrows." (Syncrude Canada Ltd.)

Exacting Standards: Kilometres of pipe and controls must be precisely assembled. A failure by a length shorter than a worker's arm in 1984 set off a disastrous fire, and a lawsuit that continues to this day. (Syncrude Canada Ltd.)

Constant Motion: Pipes for steam, pipes for hot bitumen and access roads keep Cold Lake in action, with as much of the woods preserved as possible. (Imperial Oil Ltd.)

Beats Walking: Not by coincidence alone has Fort McMurray become a hotbed of cycling – workers rediscover the efficiency and pleasure of bikes to get around the giant plant sites.
(Suncor Inc.)

ling the oil sands alone. When Syncrude started production in 1978, 11 years after Suncor, the effort was not only a product of elaborate corporate teamwork, it also included a hefty dose of government support.

Formally incorporated in 1964, the Syncrude team has a pedigree nearly as long as Suncor's in the oil sands. For years Syncrude's founder, Louisiana-based Cities Services Co., experimented with batches of oil sands products in its Lake Charles refinery, then bought Bitumount in 1958 and ran it as a pilot commercial plant processing the tarry ore at a rate of 35 tons an hour.

By this time, the "far-out think-

ing" regarding North American oil supplies (pioneered by Suncor's Pew family in the early 1940s) became common coin in an industry that had outgrown its aging reserves in the U.S. and knew it took chances by relying on the perennially unstable Middle East. At Bitumount, Cities Services was rapidly joined by others that had been prowling the oil sands and talking to Clark since the 1940s: Royalite, Gulf, Imperial and Atlantic Richfield Canada.

In the early 1960s, Syncrude looked like it would go the way of OSLO. The history that makes Fort McMurray residents look political gift horses in the mouth includes a Black Friday.

On Nov. 22, 1963, at the same time as flags flew at half mast around the globe for assassinated U.S. president John F. Kennedy, Syncrude mothballed Bitumount. Starting after Suncor, the Syncrude team lost a regulatory race to build the only big plant Alberta policy could tolerate at the time. To protect conventional oil fields that were still growing and struggling with shortages of pipeline capacity, the provincial government enforced an oil sands production ceiling. Suncor cornered the market with its 31 500 barrel-a-day plant.

Rather than give up entirely, the Syncrude team kept its Mildred Lake leases next door to Suncor's site 30 kilometres north of Fort McMurray. Syncrude launched marathon technical experiments in a research centre near Edmonton, using oil sands trucked in from the northern site via awesomely rough roads at the rate of 55 kg an hour.

The testing went on so long that it instilled staff with the patience and esprit de corps of Fort McMurray long before personnel moved there. Employee George Williamson expressed the team's mood in a wry poem preserved proudly at Syncrude (and published in a company history written by its employees, *The Syncrude Story*.

> *The future fruit of all*
> *our toil*
> *This country's huge*
> *reserves of oil*
> *But what a sad unhappy*
> *nation*
> *We used it all for*
> *experimentation.*

As the 1970s dawned, the trends that inspired Suncor and Cities Service became strong enough to light a fire under the team – and the federal and provincial governments. As the Organization of Petroleum Exporting Countries (OPEC) converted oil into a weapon of international politics by boycotting Arabia's enemies and raising prices for oil, Syncrude moved back into the regulatory arena and the field to launch construction.

Syncrude's cost estimates, following a pattern by now familiar in the oil sands, had more than doubled to $2.3 billion for a 100 000 barrel-a-day project. By the time Atlantic Richfield abandoned its 30 percent share in the project (in order to divert the monies into its share of development costs at Alaska's Prudhoe Bay), the oil sands

First Lady of the Oil Sands.
Dee Parkinson is the first woman to run an oil megaproject, because Suncor kept its practice of pioneering by placing her in charge of the oil sands division in Fort McMurray.
(Suncor Inc.)

had become a national priority under federal and provincial policies dedicated to security of energy supplies.

In epic negotiations in Edmonton and Winnipeg, fears of shortages, expectations of rising prices and ambitions for job creation overcame even the renowned political rivalry and hard feelings from fights over oil revenue-sharing between the Alberta premier, Peter Lougheed, and the prime minister, Pierre Trudeau. To fill the gap left by Atlantic Richfield, the federal government picked up a 15 percent share in the project for Petro-Canada. Ontario, as the chief energy-consuming province, took five percent of Syncrude as well as covering its bases by purchasing a 25 percent share in Suncor. On top of a 10 percent share in Syncrude, Alberta owned a half-interest in the Alberta Energy Co. (AEC). The Alberta government endowed AEC with pipeline and electric power utilities that would make money regardless of oil prices.

To put the mammoth project into production by September of 1978, a three-year construction effort set records rivalling large-scale military movements of the world wars. The labor force peaked at 10 300. The 7 000 workers in camps on the site consumed twice-weekly deliveries of 273 000 kg of groceries. A kitchen staff of 380 made six meals for the round-the-clock construction crews, who every day put away 5 000 kg of meat, 19 000 eggs, 2 000 kg of vegetables, 1 300 kg of fruit, 220 kg of coffee and 2 700 litres of milk.

With a no-strike-no-lockout agreement and internationally-competitive wages to keep the job on schedule, the army of tradesmen – and a growing minority of blue-collar women – assembled 1.1 million tons of equipment from across Canada, the U.S., Italy, France, Sweden, Germany and England.

To this day, the project continues to teach a lesson to the engineering and construction industries – every piece of such a mammoth project has to be right. The lesson goes on as of 1994, in Alberta Court of Queen's Bench, before Chief Justice Kenneth Moore. One weak link – allegedly, faulty installation of a pipe 45 cm long and 15 cm in diameter – has spawned a megaproject among lawsuits.

Six years after the plant's opening ceremonies, the pipe burst, on Aug. 15, 1984. The resulting explosion and fire shut down Syncrude for four of the last months before international oil prices collapsed.

Claims for property damage and profit losses owed to missing out on 11.7 million barrels of production exceed $1 billion. One specialty contractor alone had to collect 140 000 pages of evidence for its part in the multi-sided case, and compiling all the documentation involved in the lawsuit required creation of a new computer database.

From an outpost of 800 Natives, trappers, traders and hunters in summer and 500 in winter before the arrival of Suncor and Syncrude, Fort McMurray grew into a city of 37 000 with spectacular real-estate booms. The place proudly calls itself "Newfoundland's third-biggest city" as about one-third of the population speaks in the accents of "the Rock," after trading underemployment on Canada's rugged eastern fringe for five- and sometimes six-figure wages in the barrens of northern Alberta.

In keeping with trends of the nineties, the community christened a long hill that leads up to the plant sites from the Athabasca River valley "Supertest," in honor of the city's many bicyclists (rather than a brand of gasoline). But Fort McMurray's business is still oil. Residents, from plant managers to truck drivers, make no concessions to the outside talk about price collapses and environmental anxieties making oil a "sunset industry." No way will their industry go the way of the extinct cod fishing of offshore Newfoundland.

Despite setbacks like the explosion and fire of 1984, OSLO and the failure of a comparable giant called Alsands, oil sands production has grown. Since the mid-1980s, output has grown by almost as much as two OSLO plants. Suncor runs at nearly double its initial 31 500 barrels a day.

Syncrude has likewise nearly doubled its beginning performance, to achieve 197 000 barrels a day. With fallen prices lighting fires under the eternal oil sands hunt to reduce expenses, current production costs have been cut by as much as 50 percent, to a range of $13 per barrel for Suncor and $15 at Syncrude.

As in Suncor's pioneer days, the advances continue to rely on the cumulative effect of multiple small steps rather than spectacular breakthroughs. The process began with high-profile staff cuts and reorganizations brought on by an industry-wide price crisis that

a Syncrude executive bluntly called "a bloodbath."

In the longer term, the effort spelled more thought and work. At complexes where staff, vehicles and services attained the scale of good-sized towns, small touches added up to real money. Employees broke old habits left over from the construction boom times, like leaving company vehicles idling to keep the heaters on in winter and air-conditioners going in summer.

Retraining also became continuous as the plants reengineered and automated process and control systems. Suncor replaced hectares of first-generation computers with compact, digital systems run by a handful of staff calling up video display screens rather than patrolling long banks of tubes, valves and dials. Bigger Syncrude came to boast as much computer power as the U.S. National Aeronautics and Space Administration's launch centre at Cape Kennedy.

The innovations also go the other way, back to the humbler techniques of mining in Karl Clark's day. both plants rely steadily less on intricate bucketwheel and conveyor systems, in favor of switching to traditional mechanical shovels and trucks.

By 1994, 20 percent of Syncrude mining employed the truck-and-shovel reversion. It was combined with an innovation described as "potentially one of the biggest cost-cutting moves in years." Titled "EAPS," (Extraction Auxiliary Production System), the package uses trucks dumping into a pipeline. The pipeline was built with $12 million in scrap parts. This system shortens processing as well as driving time by mixing oil sands with water inside the trucks, on the way to the plant.

A united feeling of performing an essential national service lives on at Fort McMurray and among oil sands professionals throughout Canada – for good reason. Alberta's Energy Resources Conservation Board and Ottawa's National Energy Board say the oil sands, now about one-fifth of total Canadian production, must eventually become two-thirds or greater just to keep supplies at current levels as conventional wells run dry. Governments put money where their experts' mouths are with elaborate research efforts that, starting in the 1970s, took the work far beyond Clark's first laboratory in a U of A basement. The biggest effort, the Alberta Oil Sands Technology and Research Authority (AOSTRA), grew to a scale which qualified it for the title of "megaproject."

By the time a budget-cutting government absorbed the Crown corporation into its Department of Energy in 1994, AOSTRA had conducted about $1 billion-worth of shared-cost development work with industry, ranging from computer simulations to large-scale field tests.

By the early 1980s, government and corporate research joined to pioneer a vast new field in technology, in hopes of processing the 90 percent of the oil sands which lie too deeply buried to mine. The big step was "in-situ" production, or separating the oil from the sand deposits underground.

The work had advanced too far to be stopped by the onset of tough price controls and aggressive taxes under Ottawa's 1980 National Energy Program, or even the shaky markets and plunging prices that followed the NEP's demise in 1985.

Although Imperial dropped a $10-billion, megaproject version of in-situ production southeast of Fort McMurray at Cold Lake, the company found a way to mount large-scale development in easy stages. In his last months as Alberta's premier, Peter Lougheed called Imperial's technical and economic creativity "the best of corporate citizenship."

To avoid unbearably high initial construction costs, the company dropped plant-site processing in favor of exporting heavy-grade crude to U.S. refineries with the capacity to "upgrade" Cold Lake's molasses-like raw product.

The strategy puts conventional oil field equipment to unconventional use. Multiple wells – drilled "directionally" or at angles radiating out from compact pads to preserve the environment – inject 300°C steam and pump to the surface the resulting underground flows of heavy crude.

Diluted with higher-grade petroleum products, the hot tar is shipped by pipeline to U.S. refineries that installed upgrader equipment in the "energy-crisis" atmosphere of the 1970s and early eighties.

As Imperial profitably expanded the Cold Lake system in phases timed to match market requirements, variations on the in-situ theme proliferated. Shell Canada launched large-scale experiments in the western oil sands near Peace River. Smaller projects proliferated in between, with sponsors ranging from multinational Amoco Corp.'s Amoco Canada Petroleum Co. to home-grown Alberta Energy Co.

Industry planners crafted blueprints for a network of "stand-alone" upgraders as service centres for the oil producers. The network remains in concept stages, except for two pioneering plants built with heavy doses of federal and provincial taxpayers' money: Husky Oil Ltd.'s operation at Lloydminster and the Federated Co-op gasoline chain's installation at Regina.

Even the upgraders, much criticized by hard-core free enterprisers, took a decade or more each to go from planning into construction. Government partners, no less than the corporate sponsors, make the oil sands work by usually following the type of economic common sense which let Fort McMurray see so easily through the occasional hyperbole, such as the OSLO announcement.

A few months before that political bravado, a senior Imperial executive – refinery division chief Gordon Thomson – described how the province and the nation use the vast oil sands resource in practice:

"We can enjoy the best of all possible worlds: Using others' oil as long as they are prepared to sell it cheaply, while retaining our own valuable undeveloped resources until we can receive full value for them."

Partnership: After more than 10 years of negotiation and controversy, the federal, Alberta and Saskatchewan governments teamed up with Husky Oil Ltd. to build this industrial landmark at Lloydminster. The plant is the first of a new generation seen as necessary eventually to replace declining conventional reserves. The Biprovincial Upgrader converts molasses-like heavy crude into light oil ready for the refinery.
(Husky Oil Ltd.)

Northern Enchantment

. . . chasing visions of mammoth resources under Arctic ice

Artic Workhorse: On an August midnight on Cameron Island, a geologist packs a Beaver to head for new pastures.
(Sproule Associates Ltd.)

As oil markets wobbled towards a world price crash in 1985, Canadian business and government leaders trekked to a place renowned as only for the strong in mid-summer. Only four months earlier, the federal energy minister Pat Carney had signed an agreement with her counterparts in the Western provinces that ended government assistance for drilling on northern and offshore frontiers. Yet she led the trip to a well that oil field veterans likened to a discovery on the moon.

By the standards of a group accustomed to climate-controlled urban offices, the dress stopped just short of space suits. The party needed sunglasses, parkas, mitts and boots to ward off a dizzying combination: 24-hour-a-day sunshine reflected off ice floes and rock, amid frigid mid-summer breezes. Watchdogs – half-wolf malamutes – and armed native guards posted at discreet distances kept their eyes peeled for polar bears. All were attending to witness a 20th century industrial record akin to those set by the old sea and ice voyages to reach the North, the 19th century's versions of space programs.

Host Panarctic Oils Ltd. and president Charles Hetherington of Calgary commenced production from the Bent Horn field. This event was set in a region the old explorers had died trying to reach: on Cameron Island, halfway across the Northwest Passage, in the vicinity of the magnetic North Pole and within 1 300 kilometres of the globe's geographic top.

When Panarctic loaded the first 100 000-barrel shipment aboard ice-reinforced MV (motor vessel) *Arctic* for a 4 960-kilometre voyage to Montreal, it launched a series that continues every summer regardless of oil price gyrations. The voyages stand out as annual reminders that, like the burly Arctic watchdogs, the hunt for northern oil is a hardy hybrid.

National purpose and commercial ambition intertwine in the North. The 38-company Panarctic consortium, owned mostly by the federal government's Petro-Canada, shows this marriage of intents most clearly. But no private corporations have gone north without at least active co-operation from the federal government – and most often, the venture has been backed by generous tax incentives or outright grants.

Assertions of sovereignty over resource-rich frontiers, coupled with aspirations to raise living standards for the indigenous population, ride piggyback on the industry. These factors have been notorious in tripping up projects that tried to get ahead of the national purpose.

More than most large-scale enterprises, Canada owed its Arctic petroleum hunt to a single visionary. The prophet of the North, John Campbell (Cam) Sproule, personified the elements that matured into a multimillion-dollar marathon among industrial expeditions and captured imaginations around the globe.

Sproule had an intensely practical side. In a 40-year career as a geologist, beginning in 1927, he had a strong hand in mapping out the targets of orthodox drilling. From Turner Valley to the prolific foothills of the Rocky Mountains, including the latest extensions of industry activity into northeastern British Columbia, Sproule was actively involved. After spending 12 years with Imperial Oil Ltd., in 1951 he founded the Calgary firm Sproule Associates Ltd. It remains the senior and largest geological, engineering and energy economics consulting house. It stands out as the warehouse of experience, knowledge and technique that corporate giants hire when they want to persuade governments to let them build natural gas export facilities, or if they need appraisals to guide takeovers and asset purchases.

As a Canadian-born, independent expert, Sproule lent credibility to the often foreign-owned oil industry. In order to develop, the industry needed to build respectable cornerstones of public policies, like the idea that domestic natural-gas supplies can only be built up if exports are allowed. In the 1960s, Sproule correctly predicted that $2 billion or more would be spent to build up 200 000 to 300 000 barrels a day of production from the oil sands by the 1990s.

Sproule harnessed his education, talent and role as an elected leader in his profession's societies to a love of the North bred into him. It was his home from his birth in 1905. As the son of a travelling dentist based in Peace River, he grew up roaming a region that at the time was as remote and isolated as communities north of the Arctic Circle in the second half of the 20th century.

After starting out as a dental technician in the family practice, he became a convert to geology at age 20, when he found gold nuggets in gravel beds on a stretch of the Peace River while hiking to a fishing spot.

John Campbell (Cam) Sproule, 1905–1970, architect of the Artic oil hunt.

As he worked his way to the top of his trade – he belonged to 15 professional societies, often as a leader – he tirelessly persuaded colleagues, oil companies and Canadian governments that black gold awaited them in the North and all had a mission to go get it and civilize the place in the process! He recommended going all the way – to the Arctic Islands. The fur traders had sniffed surface seeps that hinted of Alberta's vast oil sands. Nineteenth century explorers scouted out Canada's 1.3 million square kilometres of Northern islands.

Some, like Edward Parry, had noted finding rocks that smelled of petroleum. Sproule taught that the North harbored both metallic minerals and the energy supplies that miners would need to exploit them.

They're Off. By 1969, targets were picked, equipment was in place and drilling started – here, on the Sabine Peninsula of Melville Island. (Sproule Associates Ltd.)

Right there: Surveyors mastered laying out locations on trackless ice, snow, rock and sea landscapes with few landmarks.
(Panarctic Oils Ltd.)

Official opinion agreed. Surveys begun in 1947 by the Geological Survey of Canada concluded in the 1950s that large areas of the Arctic Islands had many similarities to regions of established, large-scale petroleum production. Investors and industry had long suspected as much.

In 1920 Imperial, the Canadian arm of the U.S.-based Exxon Corp., found oil halfway along the Mackenzie River extending to the shores of the Beaufort Sea. As well, Imperial piped it through the Canol pipeline to Alaska, as part of the United States war effort in the 1940s.

To this existing technical knowledge and enticing taste of practical experience, Sproule added the scientific optimism, economic confidence, national ambition and global population projections that followed victory in the Second World War.

By the 1960s many organizations were prepared to listen to Sproule, as a dean of his profession. In repeated speeches Sproule rated the North as more than just one option. He declared "absolute necessity for Canada to strengthen her position against being overrun by exploding populations in other parts of the world. This must be done by development of all parts of Canada, including the far North, through every means possible."

In the era of space programs, industrial growth and "great and just" societies in the U.S. and Canada, Sproule drew no quarrels for suggesting such limitless possibilities as the following:

"The islands are only inaccessible on a basis of presently inadequate transportation facilities. If the incentive were present, it should not take sci-

Island Spring: Seismic crews waited only until the sun came up – and not the temperature – to go out echo-sounding for buried treasure. (Panarctic Oils Ltd.)

ence long to devise icebreakers that could move more or less at will through the Arctic Islands for the greater part of the year.

Science has accomplished so much in our generation that the development of such mechanical equipment would be classed only as a minor scientific success. Meanwhile, the idea of effective and economical large-scale submarine freight traffic is so far advanced as to be practically assured within the next five to eight years."

He described the geology as potentially so prolific that giant discoveries would make Northern oil a bargain by spreading the logistical expenses of moving it very thin over huge production volumes. He declared, "I am convinced that if geology and geophysics are properly applied in the Islands, the overall finding cost for petroleum will be substantially less than it is in the mainland areas of Western Canada."

He put his money where his mouth was. His firm constantly invested part of the bread-and-butter earnings from its orthodox industry in the Western provinces into dispatching geological field parties to the Arctic. He was fond of telling his audiences that the North could only be developed by men of vision, who are prepared to lose all when they gamble or to win riches beyond their dreams.

With that brave leadership, Sproule predicted "we could assume quite reasonably that Canada's northland will become civilized over the next 100 years to an advanced degree."

The prophet of the North died with his boots on, at the age of 65, minutes after he collapsed while reading a

Spinoff: Calgary manufacturer Canadian Foremost Ltd. grew up on specialized vehicles for the Arctic. (Panarctic Oils Otd.)

paper on Arctic geology to a conference of professionals in Jasper. But he lived long enough to launch the nation on his mission.

In 1961, two Calgary firms – Dome Petroleum and Peter Bawden Drilling – proved it was possible to work year-round within 1 600 kilometres of the North Pole, by completing a well at Winter Harbor on Melville Island. By 1967, with U.S. explorers hot on the trail of their Prudhoe Bay discovery in Alaska, Sproule persuaded the federal government to get moving in Canada's North.

The Northern development minister, Arthur Laing, formed Panarctic as a partnership with companies that had taken out vast exploration leases in the Arctic Islands but needed help to organize the collective effort and raise funds for a drilling campaign. Enthusiastic commentators described skeptics as timid souls who would rather hide their money in their socks than invest in the country's future.

From an opening commitment of $30 million that was respectable but still modest by industry standards, the Panarctic expedition rapidly escalated into an invasion. It lasted more than a decade, with periodic discoveries luring the drillers ever farther into the forbidding terrain of the off-coast islands.

Between finds, government and industry developed a doctrine that even dry holes were useful because the nation had a "need to know" the location and extent of its resources. By the time the world oil-price collapse of 1985-86 stopped Panarctic – along with most other industry activity – $750 mil-

lion had been spent to drill 174 wells.

Measured in oil, the results were disappointing: 310 million barrels but no one, big "elephant" capable of sustaining an independent production development.

The far North's wealth turned out to be natural gas: 17.3 trillion cubic feet in dry, "sweet" form that made it usable without the expensive, personally and economically hazardous processing often necessary to rid Alberta deposits of impurities.

To hunt down the resources, Panarctic pioneered a new generation of technology and methods for industrial-scale operations in severe cold over long distances. With portable housing, supplies lifted by air and sea and an attention to creature comforts considered part of the pay in the Arctic, base camp at Rea Point on Melville Island became as snug as any southern town. Rea Point became a famous eating spot, with a reputation spread far and wide by workers and visitors. Known as the "Atco Hilton" after the Calgary manufacturer of the portable housing, the camp served sumptuous, "all-you-can-eat" meals including a fresh salad bar.

Inside, it was almost possible to forget the location – except for exit signs warning against throwing doors wide open without first peeking around for stray polar bears. Eskimo dogs were allowed to roam the alleys and yards, in hopes that they would at least bark enough to alert residents of the presence of the powerful, fearless and intelligent white hunters that occupy the top of the Arctic food chain.

Soup's On: Arctic camp kitchens made up for the cold, isolation and hard labor. (Panarctic Oils Ltd.)

Brisk Work: At a location called Hecla, a rig shows Arctic conditions. Outside the weatherproofed buildings for the hands, escaping heat raises clouds of steam and flags flutter in chilling breezes. (Panarctic Oils Ltd.)

Mammoth Technology: Semi-submersible drilling caisson, a converted hulk of a Japanese supertanker, set out to drill year-round by being strong enough to resist shifting pack ice in the Beaufort Sea. (Canadian Marine Drilling Ltd.)

Warning: Off the runway at Rae Point base, this damaged DC-3 stood out as a reminder to respect northern hazards. No one was injured and the aircraft was rebuilt to fly again.

Food rode with the drilling equipment in and out of Rea Point, which doubled as a busy airport. Panarctic fielded cold-proofed rigs and outpost camps which were designed in "modules," pieces just the right size to fit snugly into aircraft. To bring a complete set for a well in it required up to 140 flights by huge Hercules cargo aircraft. As fuel costs rose with the passing years, the consortium learned to haul its equipment over custom-built ice, snow and rock roads.

In the early years, occasional accidents reminded the mechanized explorers they were up against a formidable environment. A transportation contractor's DC-3 sat on a runway, its broken wing warning new arrivals until a bigger craft hauled it away for repairs. No one was hurt in the pratfall, and the plane soon returned to service. With one catastrophic exception, the North could be a safe place to work and fly.

On Oct. 30, 1974, all but two of 34 oil field commuters aboard died in a crash of a Lockheed Electra on Melville Island. But with the blame fixed by an inquiry into pilot error, Panarctic kept flying. In the consortium's busiest period, its first 10 years, company-operated Twin Otters and Electras alone logged about 53 500 hours.

As air-transportation contractors put in long hours of heavy lifting with bigger aircraft, Panarctic also pioneered polar sea shipping on an industrial scale. In the consortium's first 10 years, vessels (chartered at a rate of up to five annually) moved 277 000 tons of supplies. Ocean shipping in the far North turned out to cost about 20 cents per kilogram of cargo, while air cargo flew for four times as much.

Panarctic rapidly became a pioneer, heavy user of satellite communications and navigational aids. Links to computers in orbit also helped develop a new drilling technique when the consortium detected drilling targets offshore of the islands. Bulky boxes of electronic gear, intended to track exact geographical positions, became standard features of ice islands that Panarctic began to manufacture in 1974.

The islands were built to support land rigs rather than bringing in costlier drillships, which were not available in tough enough form for winter work in the polar ice pack. Panarctic transplanted a technique used for making ice roads on the Mackenzie River

Chilly Passage: Even at the height of the summer open-water season, sealift freighters kept a lookout for ice and unloaded onto islands encrusted with white. (Panarctic Oils Ltd.)

and to smooth the winter skating surface of the Rideau Canal through Ottawa: holes were drilled through the ice pack and water was pumped up to the surface, to be frozen by the -40°C air into pancakes big and strong enough to hold drilling rigs.

recreated the drama of the oil blowout that showed the world the scale of Alberta's Leduc discovery 24 years earlier. Hetherington, a veteran promoter with extensive experience in the U.S. and Canadian industries before he became Panarctic's president, called the blaze a "happy accident."

Moving Day: Drilling equipment roamed on roads built from scratch of ice and rock when costlier air transport could be avoided. (Panarctic Oils Ltd.)

The drilling campaign came to an early climax in 1970, when it hit a huge gas reservoir under King Christian Island. By accident, the well blew out and caught fire. It became a spectacular beacon of success, a tower of flame, burning an estimated 50 million cubic feet of gas a day. It could be seen from a distance of 800 kilometres by airliners flying at 50 000 metres on the international polar route.

The fire, which burned unabated for three months until specialists snuffed it out in January of 1971,

"This dramatic demonstration, coupled with the earlier discovery of gas at Drake Point on Melville Island, was influential in calling world attention to the potential of the Arctic Islands as a major storehouse of energy."

Hetherington credited the spectacle with convincing U.S. corporate giants that Panarctic was on to big enough action for them to need a piece. He took their executives on a flight to see the monster flame. Soon after-

*Blazing the Trail: Huge natural gas blowout called global attention to Arctic resources and drew in international investment.
(King Christian Island, Panarctic Oils Ltd.)*

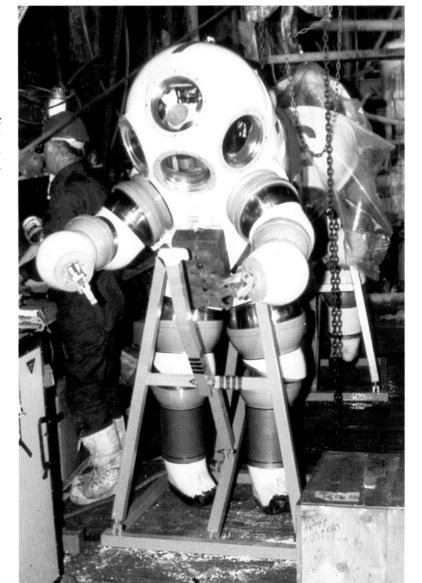

*Technology in Depth: Diver's suit, used for offshore drilling, captured the flavor of industrial exploration with space-age overtones.
(Panarctic Oils Ltd.)*

Platform Building: cold-proofed equipment worked through the Arctic night to prepare ice-drilling islands. All-terrain vehicles mounted augers to punch holes through the ice to pump up sea water and flood the natural pack until it was strong enough to support a rig. (Panarctic Oils Ltd.)

Platform Building: Like flooding a rink, making an ice island for a drilling rig took patience and gargantuan amounts of water pumped to the surface to freeze solid and thick. (Panarctic Oils Ltd.)

wards, household names of American gas transportation and production – Tenneco, Columbia, Texas Eastern and Northern Natural – committed $75 million to finance five more years of Arctic drilling. The agreements called for more money to be forthcoming from the U.S., enough for all necessary development wells, as soon as markets emerged for Arctic supplies.

Marketing projects soon emerged, on a huge scale. Polar Gas called for an immense pipeline from Melville Island to plug Arctic supplies into TransCanada PipeLines Ltd.'s domestic and export services at a junction northwest of Thunder Bay.

An optional expansion or relocation, titled "Y line," proposed a westward detour. This would meet a link drawing gas from the Mackenzie Delta, then take supplies from both northern sources across the Northwest Territories, Saskatchewan and Manitoba to the northern Ontario junction with TransCanada.

The Arctic Pilot Project tore a leaf from Sproule's book. It proposed to launch ice-proof tankers to deliver liquefied natural gas from a plant on Melville Island to southern Canada and the U.S. via the Northwest Passage, Baffin Bay, the iceberg-infested Labrador Sea and the Grand Banks of Newfoundland.

Amid gas shortages in the U.S. and fears that the Western provinces' reserves were being strained, pillars of the industry from both sides of the border backed the projects. In the late 1970s, both projects made it as far as formally applying to the Ottawa's National Energy Board and Department of Indian Affairs and Northern Development for construction and operation permits.

By this time, Panarctic was far from alone north of the Arctic Circle. Even with 800 employees in peak drilling season, Sproule's heir ranked as only a medium-sized operation by standards of the north during the "energy crisis" and price increases fostered by the Organization of Petroleum Exporting Countries in the 1970s.

Sproule's old employer, Imperial, resumed the northern hunt it had largely left alone after the Second World War and was soon joined by Gulf Canada, Shell Canada and ambitious upstart Dome on the Mackenzie Delta, then the Beaufort Sea. Promoters, reciting geological history, likened the region to the ancient deltas and shallow seas that endowed the Middle East and the coasts of the Gulf of Mexico. Prudhoe Bay, it was often observed, is not that far from Tuktoyaktuk and Inuvik by standards of an industry accustomed to think globally.

After starting out as bigger, cold-proofed but still recognizable variations of southern methods, the Delta-Beaufort drilling campaign rapidly spawned a new generation of heavy equipment and demands on personnel as it moved offshore.

Base camps at Tuktoyaktuk grew into a miniature city. Personnel commuted to the old whaling settlement by jet aircraft. Supplies arrived by air cargo and heavy shipments in barge trains pushed by tugboats down the Mackenzie River. Massive vessels were sailed or towed into the Beaufort across the top of Alaska, after crossing the Pacific, often from the giant shipyards of Japan and Korea that also built the international industry's 1960s and 1970s generations of ever-larger super-tankers.

In order to work leases in the shallows next to the Beaufort coastline, Imperial started by building islands with material dredged up from the sea floor during summer seasons of open water. The rigs worked all winter, protected by "sacrificial beaches" engineered to absorb the forces of shifting ice pack by being eroded away. Rifle-toting Native guards protected the crews from wildlife that could be as formidable as the environment.

Despite the precautions, there were hair-raising encounters with polar bears. One became legend, still talked about among veterans 20 years later. The bear tried to fight off even the bulldozer that eventually drove him away from an injured roughneck. In summer, the completely fearless bears became famous for taking swings at icebreakers if the ships inadvertently drove them off floes where they made meals of the Beaufort's abundant seals.

Rig hands and visitors, commuting around work sites by helicopters, routinely spotted the big predators. Although their white coats make them hard to see on the move, the bear are messy eaters and turn ice floes where they dine on seals red with gore.

As the industry gained experience, it adapted by evolving mammoth versions of the Native kayaks. Imperial won a governor-general's award for engineering for its flagship. It was a huge but handy steel ring named CRI,

or caisson-retained island. When empty, the ring of steel hulls floated for towing to a well site, where they were sunk to become the skeleton for a drilling island.

By 1976, to move the hunt into deeper waters, Dome sailed into the Beaufort with the nucleus of a 32-vessel, $600-million armada: Canadian Marine Drilling Ltd. With Jack Gallagher – a silver-tongued Sproule protégé – at the helm, Dome and like-minded companies that hired Canmar as a drilling contractor raised enough funds from Canadian, U.S. and Japanese investors to keep the private navy busy for nearly 10 years. Canadian governments chipped in heavily, first with tax allowances that were so generous the principal one was nicknamed "super-depletion."

Partly out of embarrassment, but more out of a desire to put Canadian-owned companies, including Petro-Canada, out in front of a hunt that it wanted to promote for "energy security" reasons, the federal government replaced tax incentives with selective grants. Under the 1980 National Energy Program, companies that proved they had the right levels of Canadian ownership received up to 80 percent of the costs of wells measured in tens of millions of dollars from the Petroleum Incentives Program.

Canmar proliferated into a self-sufficient armada, able to be frozen into the sea ice for winter at a sheltered harbor near Tuktoyaktuk called McKinley Bay. The equipment included a floating dry dock capable of lifting massive hulls and propellors for maintenance. In summer, drillships were dispersed to well sites where they were

Here Today. . . : In the shallows of the Beaufort Sea, Imperial Oil Ltd. pioneered the "sacrificial beach island" – built to withstand the shifting ice pack just long enough to drill a well, in this case called "Alerk." (Imperial Oil Ltd.)

*Ring of Steel: Tugboats towed the backbone of Imperial Oil Ltd.'s portable drilling island – CRI, or caisson-retained island – to well sites during open-water season in the Beaufort Sea.
(Imperial Oil Ltd.)*

The Neighbors: Equally at home – and formidable – on land, ice and sea. (Petro-Canada)

Sentinel: Inuvialuit like Jo-Jo Elias, as the only people with rights to shoot polar bears if necessary, were recruited to guard rig crews. (Imperial Oil Ltd.)

served by supply vessels reinforced to handle the ice. Both were protected by the world's strongest private icebreakers.

The pride of Canmar fleet, icebreaker *Kigoriak*, rode up on top of huge floes and used a spoon-shaped bow to split them by dropping through into the water. Master Clive Cunningham, a British-born veteran of the globe's oceans – like many in Canmar bridges and engine rooms – delighted in showing how the vessel's stubby strength and side-thrusting propellors let it turn on a dime.

In one display for a reporter-photographer team from Calgary, *Kigoriak* surprised a polar bear eating a seal on summer ice. The vessel outmanoeuvred the bear in a long chase, until the bear gave up running across broken floes of water and ice and turned to fight. The icebreaker beat a discreet retreat before the bear could hop up to the low main deck.

Dome and Canmar eventually adapted to Arctic conditions to a point where they could drill year-round. The Canmar engineering team's last major product – SSDC, short for semi-submersible drilling caisson – was a variation on Imperial's CRI theme, as a ship and artificial island rolled into one. Made in Japan from the hull of a 232 134-ton tanker called *World Saga*, SSDC was floated out to well sites in summer and sunk.

To keep it anchored on the dredged undersea hills, SSDC took on 193 000 tons of seawater for ballast. Precautions were taken in case the pack ice turned out to be as surly as the polar bears. With an eye on eventually learn-

ing how to build safe, permanent production installations, the industry poured research and development efforts into understanding Arctic ice. The pack slowly but constantly shifted with winds, currents, the planet's rotation and forces described as just plain "not well understood."

SSDC was built on the scale of the Pyramids and defended by space-age safety systems: 202 metres long, 53 metres wide, stiffened with 23 000 tons of steel and concrete and ringed by a 48-kilometre web of wire for underwater sensors and automatic alarms. But the seawater ballast was heated to stay liquid, so it could be pumped out to refloat the vessel in case it had to beat a retreat from the ice.

All such capabilities and more were built into the last and costliest product of Arctic ambition, Gulf Canada's $674-million Beaudril fleet. Two icebreakers as strong and nimble as *Kigoriak* – *Terry Fox* and *Kalvik* – plus reinforced supply vessels, supported a pair of drilling platforms that dwarfed even SSDC in the quest for year-round drilling, and potentially for production capabilities.

The drilling platform, Kulluk, described by a visiting American as a "giant floating hockey puck," was designed for work in deep water. Molikpaq, a portable island, worked in shallower regions.

Within two years of sailing into the Beaufort, Beaudril scored the biggest oil find in the Canadian Arctic. The well, Amauligak, flowed 5 173 barrels a day in production tests. Spectacular photographs, showing the immense flare made by burning off the test's waste products, circulated around the global oil and investment communities. An admiring Canadian Petroleum Association gave its "national journalism award" for 1984 to a Toronto account of a visit to the site via an exciting headline: that declared a new Saudi Arabia had been found in Canada's far North.

It was a long-awaited climax to the Arctic drilling campaign, after years of forecasts that peaked with predictions from Dome of 36 billion barrels of oil and 339 trillion cubic feet of natural gas. In a demonstration that Beaufort production could be practical and safe, Gulf exported tanker loads of test volumes to Japan.

MV Kigoriak is as manoeuvrable in the pack ice of the Beaufort Sea as its frequent companions in the neighborhood – polar bears. (Canadian Marine Drilling)

Amauligak turned out to be the high-water mark of northern exploration, which retreated rapidly after the discovery. Within four years, the Delta-Beaufort effort petered out, along with Panarctic's exploration. Further drilling ended with Gulf cutting its initial, 800-million-barrel estimate of its find by 40 percent.

Combined with plunging international oil prices and the abolition of NEP-PIP assistance by Carney's 1985 Western Accord on Energy, the weak follow-up results led Gulf to shelve a $4.6-billion plan to produce Amauligak at a rate exceeding 100 000 barrels a day.

The Tuktoyaktuk base camps closed. Dome sank, with debts owed to Japanese investors for money for Beaufort drilling contributing to its eventual demise in its 1988 takeover by Amoco Canada Petroleum Co. Canmar went hunting for contract work drilling for others outside the Beaufort. Beaudril was put up for sale, then went into competition with Canmar in seeking outside work when no buyers stepped forward. Political leaders in the Northwest Territories grieved loudly over losses of 3 000 jobs, and unsuccessfully appealed to Ottawa for help.

Except among critics who saw the visions of Sproule, Gallagher, Hetherington and like-minded allies as tending to inflate easily into fantasy, the Canadian North does not fade into history as a failure. Although it eventually became fashionable to quarrel about costs, the industrial explorers filled in the outlines of their maps with a respectable resource endowment.

In hindsight, looking through the gloom-tinted glasses the oil industry was forced by falling prices to don, comes a comment from the Geological Survey of Canada. Richard Procter, chief secretariat of petroleum resources appraisal, agreed "there's been a lot of disappointment." When the federal agency made a thorough count of the Northern drilling campaign's results, 26 percent was lopped off the initial forecast of oil. Gas expectations were cut by nine percent.

Yet even though the drillers fell short of finding the new Saudi Arabia in northern Canada, they added another Alberta and more to the nation's resource inventory. The GCS survey's last verdict in 1989 counted 7.1 billion barrels of oil on the Mackenzie Delta and off its shores in the Beaufort Sea – 30 percent more than its estimate of 5.4 billion barrels remaining in the Western provinces, other than the oil sands. Delta and Beaufort gas reserves added up to 68 trillion cubic feet.

The outlook for producing the oil remained cloudy because discoveries were scattered, complicated pools rather than a few giants where high output could easily pay costs of simple production and transportation systems. On one score there was never any doubt. Any production in the North was bound to be expensive. For example, in 1942, when dollars still represented gold and went far by current standards, it cost the U.S. Army Corps of Engineers $140 million to build the 1 760-kilometre Canol system.

Northern gas is no less expensive to reach, but the resource is richer. As in the Arctic Islands, the Delta and Beaufort discoveries occur in mammoth pools capable of delivering high flows requiring little processing to attain pipeline and furnace quality.

This resource wealth, while not yet at sufficient market price to integrate the north into the North American economy as foreseen by Sproule, led to a key lesson: how to fit development, when it is needed, into a sensitive natural and community environment. The prophets of Northern development promised it would be "civilizing." Natives, backed by growing environmental movements and increasingly sympathetic governments, *made* industry learn to be civil.

By 1970 Canadian oil finds, combined with ambitions to tap natural gas associated with the oil at Alaska's Prudhoe Bay, ignited pipeline planning. As prices and demand steadily rose across North America, rival proposals emerged. An international consortium that included Trans-Canada PipeLines and Canadian Arctic Gas, raced in competition against Alberta Gas Trunk Line Co. The Calgary upstart rechristened itself NOVA Corp. A nova is a star that suddenly becomes brighter. This new name expressed the scale of ambition NOVA had to grow beyond its humble birth as operator of the lines between production fields and long-distance transmission systems in Alberta. NOVA chairman Bob Blair rapidly emerged as an internationally-renowned promoter and business visionary, who in his grand designs took a back seat to no one, including Sproule.

Blair had unwavering moral support from an equally ambitious Alberta premier, Peter Lougheed.

Lougheed's government, flush with surplus royalty revenues and plans generated by oil price increases, startled even its own Conservative party by buying Pacific Western Airlines in 1974 and explaining its stock-exchange coup as a way to secure a role for Alberta as "gateway to the North."

The rivalry eventually produced a pipeline, along with hard feelings that left senior executives still not on speaking terms more than a decade later. NOVA teamed up with its counterpart in British Columbia, Westcoast Energy Inc., to construct the Foothills system. In Canadian and U.S. treaties and laws, Foothills and connected facilities south of the border remain the "prebuild" of the Alaska Natural Gas Transportation System (ANGTS).

Foothills carries nearly two billion cubic feet a day of exports from western Canada to the U.S., as a self-supporting start on the international megaproject. It has also become a mainstay of the Canadian industry.

In its first 10 years, despite saturated markets and gyrating prices, Foothills transported 2.6 trillion cubic feet of exports. This netted $9.3 billion in profits for suppliers Pan-Alberta Gas Ltd., the NOVA-controlled export dealership which became the prebuild's principal user.

Most significant Canadian gas producers belong to the supply pool. The entire nation gains from a gas project on this scale: in its first decade after commencing deliveries in 1981, the system also spun off $7 billion in federal, provincial and local government taxes and royalties. This official version of the Arctic vision has been fiercely and successfully defended against all challenges inspired by disillusionment, economic skepticism and industrial rivalry.

Criticisms have included attacks by the Independent Petroleum Association of America on the ANGTS system as a device to secure special privileges for Canadian gas in the U.S. In 1992, the Washington agency created to preserve ANGTS, the Office of the Federal Inspector, recommended scrapping the entire ANGTS structure as an obsolete "legal fiction."

Northern development ran into its greatest challenge as an immediate prospect. As the 15-company Canadian Arctic Gas consortium spent $140 million on preparations and NOVA crafted its rival project, the federal government listened to loud public questions about the future of the North and appointed a judge from the Supreme Court of British Columbia, Thomas Berger, to collect answers. After three years of hearings across southern as well as northern Canada, his inquiry in 1977 recommended a 10-year moratorium on development.

A dozen years later, back in private law practice in Vancouver, Berger dismissed as dead wrong the conventional interpretation that he sided with the era's "greens" who aggressively opposed development on preservationist principle. In his view, the environmental concerns were answered by creation of Northern Yukon National Park.

As a wildlife preserve, the park permanently sealed industry out of the "critical habitat," the caribou ranges that early versions of the pipeline routes proposed to cross to connect Prudhoe Bay and the Mackenzie Delta. Berger maintained "there was never any overriding environmental reason against a pipeline in the Mackenzie Valley."

He intended his moratorium as a way to make industrialization beneficial for the North, by blocking out time to settle Natives' land claims, letting them prepare to participate as workers and entrepreneurs and determining resource ownership plus resulting rights to royalties.

Industry got the message. The frustrated chairman of Canadian Arctic Gas, Vern Horte, eventually acknowledged Berger only did a job that had to be done: "The timing just wasn't right – you had to go through that and live with it." The painful exercise performed a service because "it made everybody more aware of things you have to consider. You had to become more careful about the environment. You had to become more careful about people." Even when industry succeeds in the North, "you have to have an environment where people work but still maintain culture, hunting, fishing and so on, even if it's almost recreational."

It took less time for the lesson to be learned than Berger expected. Less than halfway through the 10-year waiting period, Imperial and Interprovincial Pipe Line Ltd. (IPL) mounted a masterpiece of community relations to put the 500 million barrel Norman Wells oil field into production.

They did so in the midst of a population, the Dene Nation, that stood out in the 1970s as fighters of southern encroachment. (In comparison, the Natives of the Mackenzie Valley – unlike their neighbors and old enemies, the more cohesive and somewhat easier-going Inuvialuit of the Delta and Beaufort islands – have yet to be satisfied enough with the Canadian government's offers to sign the land-claims settlements.)

In fact, Imperial and IPL adapted so well that Dene chiefs boycotted the project's official opening ceremonies May 16, 1985. They stayed away not to protest the development, but to call attention to their sorrow that there was not enough work after construction ended and the project settled down to production with a staff of 100. The message was relayed by Dene political leader Richard Nerysoo, justice minister and government leader in the Northwest Territories legislature.

"There is high unemployment in the North. We have proven the capability of the Northern worker in an exceedingly harsh environment. Now we know what oil field development and pipeline construction are all about. We are prepared to participate fully in further development."

By fielding community ambassadors and bringing Northerners into the action at all practical levels, Imperial's Esso Resources and IPL built a prototype of the giant production and transportation projects planned for the Delta-Beaufort region. It has all the elements, from a long pipeline through roadless wilderness, to offshore production. Esso laid out $530 million to put 164 wells on artificial islands in the Mackenzie River, plus 130 kilometres of oil-gathering pipelines and a processing plant on the east bank of the river.

IPL spent $366 million to build a 868-kilometre pipeline from Norman Wells south to Zama, in the northwestern corner of Alberta – the farthest north terminus for the North American oil-supply system at the time.

The infant Northern community of skilled labor, service contractors and supply firms grew up fast enough, with the companies' help, to reap a healthy share of the work. Construction of the production system paid $28 million in wages to Northerners plus $90 million in service and supply orders to regional businesses. Pipeline construction employed 3 000 Northerners for two years and spun off $65 million in service and supply contracts for 250 local firms.

Within two years, it became obvious that partnership was accepted as the new rule in the North. Chevron Canada Resources Ltd. earned the first new Mackenzie Valley drilling leases granted in a decade by cutting the community of Fort Good Hope in on the action. While the four-year, $42-million drilling program had to settle for advancing geological knowledge rather than finding oil, a precedent was set in corporate adaptation that Northern leaders vowed never to let industry forget.

To win access to its targets on a 426 416-hectare block north of the Arctic circle, the subsidiary of Chevron Corp. of San Francisco formed an oil field operations joint venture with the 500-member community. Dene and Metis leaders got a say in all decisions including hiring, training, selection of oil field contractors, environmental protection and land use, plus an option to buy a 20 percent interest in any eventual production.

Chevron financed the joint venture with an initial cheque for $100 000, contributions of $50 000 annually and $400 000 in "bridging loans" for local firms to gear up with training, staff and equipment for providing oil field services. "This process is very important for the rest of the North," said the Northwest Territories economic development minister, Nick Sibbeston. "It breaks the freeze."

Confirmation arrived two years later. The north showed it had at last come to share Sproule's vision, when the industry returned to the Delta in the spring of 1989 to start picking up the pieces of the gas projects. Representatives of Imperial, Gulf and Shell Canada Ltd. trekked to Inuvik to seek answers, in hearings before a travelling panel of the National Energy Board (NEB) to vital questions for planning production. Is the Delta-Beaufort gas more than Canada needs and available for export to the market big enough to use it in sufficient quantities to warrant development? Is the North prepared to co-operate when the time comes?

The answers on both counts turned out to be "yes." The NEB awarded the companies long-term licences to export 9.2 trillion cubic feet of gas to the U.S., provided environmental screening and economic planning occurs with Northerners. Outside the hearing room on the streets of Inuvik, families egged on the industry as eagerly as leaders of job-hungry communities who testified formally.

A young Inuvialuit mother, Dinah Collins, voiced these feelings by calling the petroleum industry "hope" – for five-year-old son Tyler to have a job and for four-month-old daughter Kate to grow up into a homemaker and consumer.

At last, a maturing industry had succeeded in implanting a mood in Northerners. When markets call for the North's resources, the response will be, in a nutshell, the same as Dinah Collins's: "Go for it."

Dream Come True: Interprovincial PipeLine Ltd. at last broke the ice in the Mackenzie Valley to build a connection from Norman Wells south to the mainline's previous farthest-north site at Zama, in northwestern Alberta. (Imperial Oil Ltd.)

Barriers Fall: Mackenzie Valley native Carol McDonald showed gender as well as cultural barriers can be overcome, as a roughneck for Shehtah Drilling, a joint venture with the indigenous population of the Norman Wells region. (Imperial Oil Ltd.)

Oil Armada

. . . the invasion of Atlantic Canada

"There She Blows": Production tests in September, 1979, confirmed a major discovery at Hibernia,
on the Grand Banks of Newfoundland, by Chevron Canada Resources Ltd. and Mobil Oil Canada Ltd.
(Mobil Oil Canada Ltd.)

Talk about a gusher. Families like the Tuckers, Morrisons and Coopers changed forever after Hibernia. Chevron Canada Resources Ltd. and Mobil Oil Canada Ltd. hit this reserve of up to 650 million barrels of oil in a complicated but workable geological trap, at a forbidding but accessible spot under 80 metres of water 315 kilometres east of St. John's. In the oldest inhabited part of Canada, the hunt for oil beneath the Grand Banks of Newfoundland spawned personal discoveries of no less significance.

Caleb Tucker explains in the self-effacing manner which is a trademark of outports like his home Quidi Vidi Village. He tells of a friend who put some of his $9 000 a year income from cod fishing and unemployment insurance into a training course, a helmet and steel-toed boots, then quadrupled his income by going to sea on a drilling rig. "My buddy found out what he was worth."

A generation of drilling, which culminated in discoveries offshore Nova Scotia at Sable Island as well as on the Grand Banks, raised a tide of expectations. Hibernia made waves when it was discovered in 1979 and swelled hopes by making them feel realistic. The East Coast was an old hunting ground by then – almost as old as Alberta.

In 1944, driven by wartime hunger for fuel and the outlook for an increasingly motorized society when the soldiers came home, oil prospectors prowled nearly two million hectares of mineral leases in Nova Scotia and Prince Edward Island. The area under exploration placed second only to the vast tracts that the industry scoured in Alberta during the hunt that finally found Leduc in 1947. Mobil, pioneer believer in the East Coast in the same way that Suncor became the leading early custodian of hopes for Alberta's oil sands, drilled Canada's first offshore well in 1943. Subsidiary Island Development Co. built an island of wood and rock in eight metres of water 13 kilometres offshore of Prince Edward Island, and spent $1.25 million on a well 4 479 metres deep called "Hillsborough No. 1." It was dry, yet anything but a dead end.

As world oil use steadily multiplied more than five-fold between 1948 and 1972, early failures only goaded the "explorationists," who then held most of the industry's top jobs, to look farther and harder. By 1958, Mobil returned to Atlantic Canada. It alone took out 445 000 hectares of offshore leases,

The Beginning: Mobil Oil Canada Ltd. drilled the first Canadian offshore well in 1943.
(Mobil Oil Canada Ltd.)

setting off a stampede that by 1975 staked out 1.3 million square kilometres of ocean floors.

As the global hunt accelerated, Atlantic Canadians like Cliff Morrison became living versions of statistical economic indicators. Born on Cape Breton Island – celebrated in the region as second only to Newfoundland for generations of hard times – he heard about fortunes associated with oil as a school boy in Halifax. He had a first-hand taste in Calgary in 1974. A visit for a wedding stretched into a summer job stocking shelves in a new department store built in the retail offshoot of Alberta's oil and real-estate boom.

Inspired to aim higher on the job ladder, he earned a trade ticket from a technical school in Halifax as a drafts-man. Returning to Alberta, he went from draftsman to oil field roughneck to construction contractor, raising his income each step of the way. As the West headed into the slump brought on by the 1980 National Energy Program and double-digit interest rates, the still-accelerating oil hunt on the East Coast came to the rescue. Back in Dartmouth, Morrison and wife Murreen, a nurse, found good jobs and bought a house in a suburb nurtured by the economic activity that offshore drilling spawned throughout the region.

In 1985, the family returned to Alberta because of a flurry of work brought about by the dismantling of the NEP's strict price regulation, sti-fling export controls and crippling taxes by the Western Accord on Energy. After the 1986 world oil-price crash cut short the Western revival, the family, which by this time included two children, headed back East to get

in line for a share of better times fostered there by the 1988 government and industry agreement on building a $5.2-billion production system for Hi-bernia.

All the bouncing around became normal for a generation that, as the Tuckers of Newfoundland observed, learned at the hands of the petroleum industry it could do better than the old, seasonal treadmill of fishing and unemployment. Before the recession triggered by the NEP took hold, Atlantic Canadians like Morrison marched out to Alberta at a rate of nearly 15 000 a year.

Then 22 000 headed back to the East Coast as the oil hunt peaked there, while the West hit bottom in 1982 – 84. The 5 000-kilometre treks – often done the hard way, with belongings crammed into pickup trucks and rented trailers – were driven by a faith that was new to a region celebrated for being hopelessly down on its luck since the decline of sailing ships. The belief is expressed by friends and kindred spirits of the Morrisons, such as Cathy Cooper, whose husband Carl came from a Cape Breton family.

The Coopers had sent 8 of their 10 children off to the migrations fostered by the shifting oil hunt. Cathy Cooper travelled with her husband, sharing an unshakable belief: "If you have your stuff together and want to work, you can find a job."

Hardly noticed by outsiders, oil spelled a cultural revolution in Atlantic Canada. Russell MacClellan, the Liberal MP for Cape Breton, who served as the party's energy critic in opposition in the 1980s, described the effect oil had

Old Port, New Home: Drillships found a welcome port in the narrow entrance to the harbor at St. John's, like generations of fishing and war fleets before them. (Mobil Oil Canada Ltd.)

Stilts: In shallow water – here, a well into the Venture field of Nova Scotia by Mobil Oil Canada Ltd. – "jack-up" rigs parked on the sea floor on steel legs. (Petro-Canada)

"Drillings's Drilling": Except for the size of some of the hardware and the slowly rocking deck, a rig floor offshore differed little from counterparts on land. (Husky Oil Ltd.)

On the Bridge: Informal dress prevailed on offshore rig, where high levels of technical expertise were employed in the comfort of denim. (Petro-Canada)

in the same language as foreign-aid workers used to explain the hard part of their jobs – imparting motivation to Third World communities. On a national tour during the bottom of the 1986 oil-price collapse, MacClellan observed he could see the difference the industry made even then.

In areas accustomed to oil boom 'n bust cycles, technical, professional and managerial personnel took lean times as signals to try harder. After layoffs they kept their hopes alive and often started new companies from reorganized old ones while adapting to the causes of the slumps.

Even at their worst, MacClellan rated communities wounded by the oil-price crash as hotbeds of optimism, ambition and activity compared to parts of his constituency. He described Sydney, N.S., as "stifled by decades of 40 percent unemployment." Rapidly, MacClellan became a political convert to encouraging oil activity.

Work proliferated on the Scotian Shelf and the Grand Banks through the 1970s and early eighties. Hibernia stood out as the high-water mark of offshore drilling. The East Coast often lived up to advance billing given to it by geologists. Using a combination of progressive seismic-survey technology and theories born in an earth-sciences renaissance of the 1960s, geologists guided explorations to a string of successes.

The new view considered Canada's Eastern Continental shelf to be a cousin to the oil-rich floor of the North Sea between Britain and Europe. The multibillion-year narrative of "plate tectonics" describes the conti-

nents as crusts in motion atop global formations that generate titanic forces of pressure and heat.

According to this theory, the Scotian Shelf, Grand Banks and Labrador Shelf started as parts of the same basin as in the North Sea. The basin evolved into the Atlantic Ocean as an ancient giant land mass split into Europe, Africa and the Americas.

As the continents drifted apart, the shallow shelves off eastern Canada collected eroded sediments which form traps to pressure-cook dead flora and fauna into oil and natural gas.

Drilling rigs soon proved geologists were on the right track. Shell Canada Ltd. (the nation's top gas producer until Amoco Canada Petroleum Co. jumped ahead by buying Dome Petroleum Ltd. with its vast Western reserves in 1988) found itself right at home on the Scotian Shelf. In the first well off Nova Scotia in 1967 at Sable Island, Shell found gas. It was at higher pressures than the drilling equipment of the time could handle, and the well had to be abandoned.

Even with improved technology, a combination of extreme pressures and intricate geological faulting turned the well called "West Venture" into a marathon exercise in controlling a subsea blowout that eventually cost $167 million. Six years before the discovery of Hibernia on the Grand Banks, Mobil found oil off Nova Scotia, in 1973, with a well called "Cohasset."

The federal government urged on the hunt. The 1970s "energy crisis" developed out of rising consumer demand. This overtook established North American oil fields which were being depleted. As well, there were price increases and politically-inspired supply boycotts by the Organization of Petroleum Exporting Countries (OPEC).

In addition to the lure of early successes and geological predictions that there was more in federally-controlled offshore and Arctic oil frontiers, were the Liberal government's added tax incentives, resource leasing terms and a mandate to grow for the Crown energy company Petro-Canada.

In 1980 the NEP accelerated the offshore hunt. As well, it added highly political wrinkles inspired by popular demand for "energy security" and suspicion of international oil companies, as prices peaked in the wake of the 1979 Iranian revolution.

A provision known as the "Crown back-in" – and deplored by some industry veterans as little short of retroactive nationalization of discoveries – reserved rights for the federal government to ensure a minimum of 25 percent Canadian ownership for offshore supplies. Nationality likewise became a key factor in the flow of money into seabed exploration. The NEP's Petroleum Incentives Program (PIP) paid up to 80 percent of exploration costs on federally-controlled prospects but reserved the top benefits for Canadian-owned companies.

Less than one-third as much went to corporations with foreign owners, including the pioneers of offshore drilling. This economic nationalism had an extra hard edge as a result of the way the NEP raised money to pay for PIP.

There was a "redistributive" side. The federal government played Robin Hood with the Petroleum and Gas Revenue Tax (PGRT). It financed PIP by taking up to a 16 percent share of gross sales revenues from established reserves, which were dominated by the senior, often foreign-controlled companies.

Petro-Canada, labelling the rigs it hired with its trademark maple leaf, sailed to the top of offshore activity when total spending off the East Coast swelled to $5 billion between the NEP's introduction and its abolition in 1985 by the Western Accord on Energy. But the national oil company had no monopoly on high expectations.

Two pillars of the Alberta industry – Husky Oil Ltd. and Bow Valley Industries Ltd. – went to sea at a pace that matched Petro-Canada. The pair launched a corporate version of Atlantic Canadians' continent-spanning treks with "Bow Drill," a partnership that became a household word on the East Coast.

When provincially-chartered NOVA Corp. bought Husky from its Wyoming founders in a 1970s expansion drive cheered on by Alberta political leaders, the combination had good business reasons to test new waters. It needed the kind of oil found offshore.

About three-quarters of Husky's reserves were molasses-like heavy crude that needed expensive processing or "upgrading" to be used by Canadian refineries. This awkward legacy dated back to a decision to concentrate on resource leases surrounding Lloydminster on the Alberta-Saskatchewan boundary in the exploration rush after the 1947 Leduc find.

Pendulum Ride: Atlantic swells rocked even mammoth drilling rigs, making the standard method of transferring between vessels an exhilirating few moments. (Husky Oil Ltd.)

Sea Monster: New generation of "semi-submersible" drilling rigs were floating villages that dwarfed land equipment. (Petro-Canada)

The takeover brought Husky into a corporate family celebrated for visionary ambition. At the time this included seizing a leading role in international projects to open up the Arctic with natural gas pipelines. In the late 1970s and early eighties, the NOVA family had energy to spare – even from the northern pipeline rivalry and marathon negotiations that eventually drew the federal, Alberta and Saskatchewan governments into a partnership with Husky on an upgrader plant at Lloydminster.

A willing partner for offshore enterprise stood out, just a stroll across downtown Calgary from Husky's headquarters. He certainly had the right connections and experience. He even had an appropriate name. Bow Valley's founder, Daryl K. (Doc) Seaman, rubbed shoulders with Canada's financial and political elite. He served with former Liberal finance minister Donald Macdonald as the sole businessman on a national royal commission into the economy and Canadian unity (which eventually produced the recommendations that the Conservative government of Brian Mulroney translated into the 1987 free-trade agreement with the United States).

Seaman's investment backers in Bow Valley included the Montreal branch of the Bronfman family, as Cemp Investments Ltd., plus the nation's biggest credit union, Caisse de depot et placement du Quebec. The cornerstones of Bow Valley's growth beyond its 1950s and sixties status as a healthy but small Alberta firm were a string of offshore discoveries in the North Sea, on leases acquired in a takeover.

Seaman personified a strong strain of adventurous grit that often showed through the corporate polish and technical university credentials of his generation. NOVA chairman Bob Blair wore his cowboy boots to meet Toronto financiers just as proudly as he wore the name "Iron Horse," given to him by the Natives affected by his pipeline work. And Seaman stayed as down to earth as the rig hands from St. John's and Cape Breton who worked for him.

He liked to be called "Doc" by all, from office clerks to business barons. He earned the nickname by carrying his gear in a black leather satchel to sandlot baseball games, which he played for better money than many jobs paid around his hometown of Rouleau, Sask.

He had learned to take calculated risks where the stakes were life itself, as a bomber pilot over Europe during the Second World War. After leg wounds cut short a promising hockey career, he picked up a degree in engineering in Regina and headed out to Alberta "on spec" within months of the Leduc discovery.

After nearly 40 years of parlaying a $5 000 investment in a seismic exploration drill into a 3 000-employee, international oil empire later, in the mid-1970s he let the stuff of company builders show. How could he bring himself, in effect, to "bet the company" on Bow Valley's first North Sea well? It was a pure "wildcat," industry jargon for first wells drilled into untried geological formations. It cost as much as Bow Valley's entire annual cash flow at the time.

"You just suck up your guts and

go for it," Seaman said. Geological science, engineering and financial expertise help calculate risks and turn profits, but finding new oil supplies of any size "takes more of a venturesome spirit. It's the nature of the business. You take commensurately more risk to make a larger discovery. If you stay within known areas, you're apt to be restricted in what you find," said Seaman.

The Bow Drill partners went for it on the Grand Banks and Scotian Shelf with a $400-million armada, led by three deep-water drilling vessels called "semi-submersibles." This was last-generation, high technology. As the industry headed ever farther out to sea, breakwater-like islands of stone and wood and drillships evolved into "jack-up" platforms.

For towing out to work sites, these semi-submersibles floated high in the water. They were parked atop well sites, then the steel legs were filled with water ballast to lower and stabilize them. Arrays of anchors, engines, propellors and "dynamic positioning" helped pinpoint locations, in conjunction with employed continuous communications with navigational aids aboard space satellites. Support vessels ranged from miniature submarines with robot arms for working with seabed drilling equipment to specially-built supply ships that cost an estimated $20 000 a day each to operate at the height of their demand.

On a personal scale, rig hands risked no less. In case anyone had forgotten the monuments to fishermen lost at sea that add a

sombre note to parks in many East Coast communities, tragedy struck again in 1982. The Grand Banks also became the grave of a rig called the Ocean Ranger, which went down in a storm at a cost of 84 lives.

Each of the three Bow Drills incorporated about $10 million-worth of improvements intended to prevent a repetition of the disaster. And it was not repeated in Canadian waters as industry activity peaked in the mid-1980s, with more than a dozen rigs and navies of support vessels operating offshore Newfoundland and Nova Scotia.

Standard features came to include: fast-acting systems for preventing "listing" by countering water ballast shifts in high seas; back-up control networks; enclosed lifeboats; elaborate personal safety gear; and extensive training on shore for all staff.

No one took the long helicopter rides out to the Grand Banks rigs – which were considered among the most hazardous parts of the work –

Cold Comfort: A full-length, inflatable safety suit was the minimum necessary to survive more than a few minutes in cold northern waters; the flippers on this one were a rare option. (Petro-Canada)

without sweltering in rubberized, water-tight, bright orange safety suits. Worth about $800 each, they were one-man rafts that inflated in seconds, kept out the deadly chill of water that barely rises above freezing even in summer, and included lights and whistles to guide rescuers. Although rig hands took off the suits to work, lockers full of them were kept within easy reach of all job locations.

Like Seaman, the personnel took their chances with open eyes. The rigs collected adventurous spirits. Author Farley Mowat – a Cape Breton resident who spent months tossing about in a tugboat on the North Atlantic for his classic book about the salvage business, *Grey Seas Under* (Toronto: Little, Brown & Co., 1958) – described the oil-drilling fleet as a bigger revival of 19th-century whaling under a gloss of high technology.

One spill would kill as much wildlife as generations of Atlantic Canadians who extracted lamp fuel and lubricants from whales, seal and walrus, Mowat observed. In his eyes, the Ocean Ranger disaster only served as a reminder of how technological pride has gone before falls in the region, in cases dating back to the sinking of the unsinkable *Titanic*:

"There's never been any fail-safe system for the North Atlantic. The idea we could ever conquer the ocean is ludicrous."

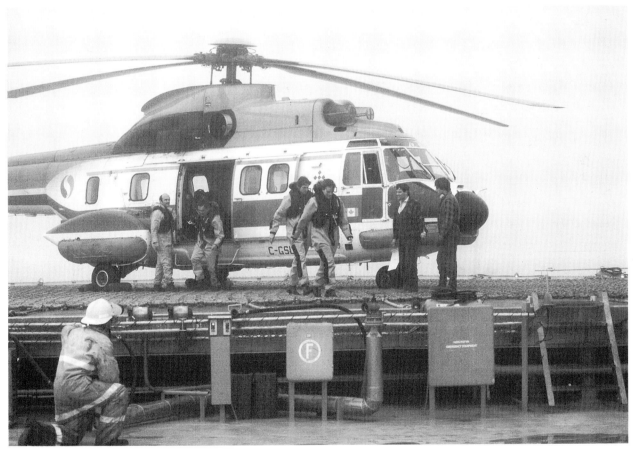

Instructive Welcome: The first sight on arriving aboard an offshore rig in mandatory flotation suits, was a crouching firefighter ready for instant action – as an unforgettable reminder that this was no job for the careless. (Husky Oil Ltd.)

Mowat concluded that with measures to prevent new disasters based on lessons from the last one:

"We can reduce the odds, and that's about all."

Mowat drew no quarrels from hands like John Morrier. As a deep-sea diver, Morrier used some of the highest technology in the fleet by operating miniature submarines. With pride, he acknowledged the chances he took by going down 90 metres underwater or more while lying prone in a cigar-shaped vessel too small for a circus contortionist to turn around inside of, much less allow the occupant struggle into one of the orange rescue suits in an emergency.

"It's the most risky job you can do here. Chances of surviving if anything goes wrong are virtually nil."

The native of Drummondville, Que., explained why he trained and competed for this dubious privilege, in sentiments that would have been at home in the old whaling memoirs:

"I had a nine-to-five city job – and I was bored to tears."

An expedition mood of united concentration on a formidable task pervaded all ranks. The purpose of the work created an undertone of hunting for buried treasure. Bow Drill chalked up a string of discoveries awaiting commercial development, in turn setting off big movements in stock ex-

Perfect Shape: Specialists maintained miniature submarines with painstaking care. Beside being among the most expensive pieces of offshore equipment, they rated as the most hazardous because there was little or no chance of escape if something went wrong at work, 100 metres deep in the chill Atlantic.
(Petro-Canada)

changes, with results such as North Ben Nevis, which delivered 6 000 barrels a day in production tests and was estimated as capable of more than double that.

For the crews that worked towards such finds, the sea provided inspiration to stay in the hunt – to the point of addiction.

"Drilling's drilling," said George Fukas, a 27-year veteran of western oil fields who went to sea as a rig superintendent. "But I love life on an offshore drilling rig."

The working conditions helped. Hands served rotations of 21 days of pure labor interrupted only by meals celebrated for their generosity and quality, alternating with three-week holidays.

"Who else," said Fukas, "works six months, gets six months off and gets full pay?"

The hazards welded the crews together, and made them part of a wider maritime community.

"Everybody helps each other at sea," said drilling supervisor Bob Straub, reciting how an oil-rig supply ship rescued the crew of a fishing boat sinking in bad weather off Nova Scotia. Offshore rigs had a way everywhere of exposing human nature at its best. Fukas had his first revelation as a beginner aboard one in the North Sea.

"Everybody thought they'd had it," in a howling gale that whipped up seven-storey waves and sank another rig nearby. He was struck by the unanimous response – calm determination. "Everybody became concerned for the next guy. Everybody was asking each

other, 'are you all right?'"

At the corporate level, a similar spirit spread as the industry learned more about the North Atlantic environment. Officially, companies were far more optimistic about technology than Mowat. In practice, all recognized the region's challenges and risks. A co-operative to prevent and clean up spills came together.

Collective resources were poured into a new safety science called "ice management." Ice abounds offshore of Newfoundland, even in late spring and early summer. Just a "bergy bit," as the fishermen called the leftover stump of a mostly melted iceberg, could pose a threat to the biggest rig. Even a modest chunk packs a wallop of 6 000 tons or more. The steel rigs could withstand the first blow or two as such a chunk arrived at its drift speed of one to two kilometres an hour.

No equipment could tolerate repeated smashings from a bergy bit caught in the platform's structure and tossed against it, by a sea that Newfoundlanders rated as calm when the perpetual swell ran two metres high and slowly rocked even the biggest rig.

The ice made a big contribution to costs that hit an average $63.2 million per well during the offshore drilling peak of 1981-85. In any given 24-hour period, a rig on the Grand Banks was almost certain to spend considerable time disconnecting from its well, weighing anchor and propelling itself out of the way of icebergs. The shuffling around generated some of the most hazardous work.

The rigs' anchors, for example, were too big just to be winched up.

Supply ships helped by grappling the chains. As the vessels took up the strain, steel cables stretched taut as bowstrings. Crews scrambled off the open rear decks into sheltered spots forward.

At such moments, a supply ship had only to dip its stern into a two-metre wave trough for the wire rope to pop up from between two guide posts and whip across the deck in the blink of an eye. The dodging deckhands and manoeuvring rigs followed a rule of thumb. This rule, articulated by a Bow Drill captain, Terry Robinson, was, "You've got to have a lot of respect for the ice and the Murphy factor."

The same went for upper management. Manager of the Husky-Bow Valley offshore partnership, Larry Prather, observed, "We tend to forget we're here to drill for oil and gas and find ourselves avoiding icebergs." The

floating ice often forced each rig to keep three supply ships around, rather than the two required in clear seas.

Collecting radar, satellite and lookout observations, the industry ran an ice-watch network that also drew on specialists in academic and government agencies. The system labelled all the bergs and bits with numbers, charted them, relayed sightings to all the rigs and attempted to understand the interplay of wind, waves, currents and ice shapes well enough to predict movements by the hazards. Rather than dodge every bergy bit with the time-consuming, expensive disconnection and reconnection procedures, the industry set out to master at least the smaller floating menaces.

A favorite technique combined the cowboy lasso and the fishing net. Supply ships sailed close enough to ice chunks to sling rope around them. The

Making Fast: A supply ship's crew prepares it to tow an iceberg. (Petro-Canada)

ice was too smooth for rope to hold, so engineers added snares made of nets 72 metres long and 18 metres from top to bottom, with mesh 3.6 metres across. The supply ships powered up their diesel engines for tug-of-wars lasting hours, to tow or at least divert the bergs away from drilling rigs. They met with varying luck, depending on waves, winds, currents and the ice shapes they played upon. In the oil-drilling fleet, the sailors' "make 'n mend" chores included long hours repairing monumental frays and tangles in the leg-thick ropes used to try harnessing the ice.

In all cases, drilling operations did research and field testing with an eye on developing much larger-scale uses for new techniques when the time came to produce their discoveries. Mobil even made a motion picture about matching steel against ice, with touches of wry humor and out-takes from home movies. The film fit in well with the plain-spoken population on the oil frontiers.

To test the forces and materials involved in planning a production system for Hibernia, the company made truck-sized battering rams and instrument arrays in Calgary, took them north and pounded icebergs at Pond Inlet before audiences of delighted children.

In the mid-1980s Arctic and Atlantic offshore drilling peaked simultaneously. Experimentation and design work reached a point where Calgary boasted one of the biggest Canadian chapters of SNAME, the international Society of Naval Architects and Marine Engineers. More than 90 professionals belonged, working for oil or allied consulting firms on projects ranging from designing jumbo icebreaking tankers to a radar system tailored for tracking

Full Ahead: A supply ship attempts to tow an iceberg – looking mountainous but ordinary by Grand Banks standards – off a collision course with a drilling rig. (Petro-Canada)

iceberg movements.

Smaller but still active, the group remains a hotbed of expectations for ultimate development of oil and gas under Canada's seabeds. Members often point out that mixed results of an undertaking on the scale of offshore exploration do not spell failure. It took more than 200 wells for world industry to find oil in the North Sea.

By 1990, long after depressed oil and gas prices virtually sank ocean drilling, the total number of wells drilled in offshore Atlantic Canada was still only 254. Reserves forecasts remain guesswork based on limited information by geological and engineering standards.

Hibernia was the 138th well of offshore Atlantic Canada, but eventually turned out not to be the first commercial discovery. That honor goes to

Mobil's offshore find Cohasset, near Nova Scotia. In combination with follow-up drilling by Petro-Canada and a further find nearby of Shell's called "Panuke," the field has now entered production while Hibernia remains under construction.

At Panuke-Cohasset, Crown corporation Nova Scotia Resources Ltd. teamed up with British-based LASMO (London & Scottish Marine Oil, a North Sea veteran) to pioneer offshore development on a budget. The production system employs a converted exploration rig as a production platform, to fill small tankers with a high grade of oil prized by refineries.

A larger but similarly floating production system is under development by Petro-Canada for the second-richest find so far after Hibernia, Terra Nova. With Husky as a minority silent partner, Petro-Canada's executives

Contender: Production testing Terra Nova, a find by Petro-Canada, which has vowed to make it the next production development on the Grand Banks. (Petro-Canada)

confidently told every shareholders' meeting held since its partial "privatization" sale in 1991 that the long offshore drilling campaign will pay off handsomely. Tapping the resources off Atlantic Canada stands out as a top priority for the national energy enterprise. Hibernia, in this strategy, is described as laying cornerstones for further development by establishing an "infrastructure" of services and skilled labor. And not a moment too soon, as effects of 1990s conservation bans on cod fishing ripple through Newfoundland's traditional economy.

In other corners of the fledgling offshore oil industry, the depressing effects of the 1986 oil-price collapse were amplified by the dismantling of the NEP. PIP died too, with no signifi-

cant replacements. The NEP's nemesis, the Western Accord on Energy, made revolutionary changes in Canadian policy when viewed from the perspective of oil and gas frontiers.

For the first time in generations, the Accord enacted a policy of letting economic forces call the tune for development. Energy security has been largely entrusted to an increasingly open market fostered by free trade in gas and oil. Primarily trade is with the U.S., but also with overseas oil suppliers led by North Sea production that pours by tanker-loads into Montreal via the St. Lawrence Seaway.

Rather than disillusionment with the resource endowment, hard economic realities – led by prices – re-

Flaming Bounty: A fireman beams at the sight of proof that the Grand Banks harbor oil wealth, as he stands by in case of any accidents during production testing of Petro-Canada's Terra Nova discovery.
(Petro-Canada)

main mostly responsible for largely shutting down exploration on the Grand Banks and Scotian Shelf since the mid-1980s. With the blunt candor that Atlantic Canada has come to expect, and respect, in discussions development and employment prospects, Shell's president at the time, Jack MacLeod, explained:

"In respect of natural gas resources, there is nothing concerning provincial, federal or private policies and strategies that is inappropriate in the sense of being an inhibiting influence on development at present. The inhibiting influence is the fact that reserves discovered to date will not support economic development in present market conditions."

He made no false promises, saying only that development will come "at some future date when the economics are right . . . that date cannot now be foreseen clearly."

By "economic," MacLeod's generation of oilmen had learned to mean more than just profits for companies. Benefits to the country have to be taken into account in planning projects, because they cannot expect to escape sharing their gains with the rest of the population. MacLeod stressed that, to be justified, a development must "yield meaningful royalty and tax revenues to the Canadians who own the resources through their governments, and an adequate return to the investors of development capital."

Tapping the gas offshore of Nova Scotia, with an ambitious plan known as Venture, would be almost as tall an order as the Hibernia megaproject. Production and pipeline systems to produce and deliver 400 million cubic feet a day of gas would cost an estimated $4.4 billion. The expense would be $1.30 per thousand cubic feet of reserves tapped by the project.

Sales prices fell lower in the gas slump that came in the late 1980s on the heels of the oil collapse. Also in sharp contrast, the unit-cost involved in Shell's central-Alberta Caroline project was only about 30 cents per thousand cubic feet, and the gas there yielded rich streams of liquid-fuel and sulphur by-products.

But even with ex-Nova Scotian MacLeod retired, the gas side of the petroleum industry is not forgetting entirely about its discoveries in Atlantic Canada. Small pipeline projects continue to probe markets in New England, at this early stage with long-distance shipments of gas from western Canada. These are seen as toeholds for bigger things in future. MacLeod, who started the projects, explained:

"By establishing a modest presence in this market, we will contribute to building customer confidence for Canadian gas. This should be helpful one day, when larger volumes become available from eastern Canada."

The Conservative federal energy ministers, Marcel Masse and Jake Epp, who negotiated for construction of Hibernia's 110 000 barrel-a-day production system, acknowledged the arrangements added up to a huge exception to the free-market policy.

To prod the project consortium to go to work, Epp eventually had to go beyond initial 1988 federal commitments to cover 60 percent of construction costs with $3 billion in grants, loan guarantees, interest-rate subsidies and temporary financing.

Mobil, Chevron and Petro-Canada required more help when Gulf Canada, expanding into Russia amid poor oil and gas prices, abandoned its 25 percent share of Hibernia in February of 1992. By shopping the vacated share around, a lone taker was found, Murphy Oil Co. of Arkansas. Murphy bought only 6.5 percent, so the federal government took on 8.5 percent ownership for $290 million.

Epp, like all his predecessors who worked for a Hibernia production project since the discovery in 1979, insists much more was being done than to create an annual average 2 200 construction jobs for five years, then 1 100 permanent positions for operators.

The project founds "a new petroleum province," with a chance of eventually helping to pull Atlantic Canada out of its legendary hard times

the same way the industry cleared away lingering effects of the 1930s Great Depression from Alberta.

While the industry harbors no shortage of skepticism about when energy prices might recover to the point where offshore reserves look attractive on a purely commercial basis, it is virtually impossible to find a leader who begrudges the help to Newfoundland.

Even during the worst of the 1986 oil-price collapse, when memories of NEP-PIP-PGRT were still fresh, the voice of the senior oil companies who were hurt most by economic nationalism made no attempt to intervene in discussions of aid for the Newfoundland project.

The Canadian Petroleum Association's chairman during the crisis, Shell executive vice-president Doug Stoneman, invoked the ancient moral injunction against beggaring thy neighbors.

Thousand-yard Stare: Long shifts handling heavy equipment and high responsibility for expensive, hazardous operations show on the faces of seagoing oilmen like Petro-Canada rig representative Peter Moew. (Petro-Canada)

Born Again

. . . in conclusion, new beginnings

They called it "no fun at all," but breadwinners like Gerard Butts, Sam McCollum and Todd Knight unwillingly played a role in generating a reverse pride in the "propulsive industry," as petroleum's leaders declared the expansionist, "exploration phase" to be over. The engine of growth stalled after energy prices fell, setting off two waves of austerity that cost an estimated 25 000 jobs in the first year (after the 1986 oil price collapse), then another 12 000 as natural gas prices tumbled.

As "restructuring" dragged on, with corporate mergers and asset sales compounding employment losses, casualties like Butts, McCollum and Knight came to stand for two propositions. The change, described by Amoco Chairman T. Don Stacy as nothing less than "a bloodless revolution," spared no one. This backlash response earned oil headquarters, the city of Calgary, a reputation as a capital of entrepreneurship. Long after the industry's free-spending boom times ended, petroleum left a legacy of drive and optimism.

McCollum had no shortage of material for despair. When he lost his job in middle management, he was a 46-year-old father of seven and pillar of the Mormons who had risen beyond the rank of bishop to senior church adviser. The manner of his loss was typical of corporate policies that favored clean breaks.

"I had a staff of 14 people and everything going for me. I didn't do anything wrong. I went to work one morning and my vice-president laid me off. I was out of a job in a moment's notice."

Yet within months, he declared "I got lucky."

Rather than send out résumés to other oil employers who had adopted hiring freezes at best and staff-cutting programs at worst, he set up a family company with wife Jeannine and son Cliff. They figured out a way to turn austerity into a new business niche, where they marketed cost-cutting methods of performing McCollum's specialty in corporate data management.

Butts, a father of five also in his 40s, likewise had all the makings of a victim. Within 18 months, he lost his job with a six-digit salary running a 44-employee corporate department, sold an acreage home, parted with the family's prized horses, declared personal bankruptcy and endured a heart attack. No churchman, he said "everything went to Hell."

Yet his 20-year oil career taught him to keep counting on the self-reliance that sent his ancestors to western Canada. He rose to executive rank by devising a computer system to manage oil companies' land portfolios, building a consulting firm on the package, selling that to a large corporation, then hiring on with the buyer.

He made the family hobbies – wood and leather-working, plus horses – into a business by manufacturing and marketing a line of accessories in tack and saddlery shops. His wife Peggy, a descendent of homesteaders who arrived in Alberta a year after it became a province in 1905, voiced the hardy, driving optimism that helped proliferate small businesses after the oil crash:

"I keep seeing the rainbow at the end of the storm. I just say, we can do it."

Knight also drew on his hobbies. They took him back into the petroleum industry as part of its officer class, with scarcely any absence after he lost his rank-and-file job as a geophysicist.

As a university student, he had become an avid "hacker" at a time when personal computers were still new in the 1970s. He made himself a programmer by learning to do class assignments on his PC. After his job loss, he hauled out the self-taught skill to create a specialty firm that married computers and geophysics for rapid, cost-saving and more accurate interpretation of seismic-survey results. Within four years, he had a staff of seven in plush downtown Calgary office quarters, $1.5 million-worth of computer gear and strong growth prospects fostered by a steadily lengthening client list. When mixed with the right combination of technical skills and tastes of success, dismissal became a catalyst for advancement, rather than a lasting defeat. Knight, describing his own experience, voiced the creed of a new oil generation whose companies became stock exchange darlings in the early 1990s.

"It seems like every entrepreneur has some bad situation in his past. They were laid off, fired, went through a divorce or had some other traumatic knock. You go through stages. There's shock – you didn't believe it could happen to you. There's an anger phase. Then comes the rationalization. You work out that you've got to get on with life and make it all happen."

The industry went through the same stages. Initial shock and disbelief over the collapse of oil prices, from nearly $35 US per barrel to $10 in early 1986, spawned a generation of economic projection that later came to be called the "hockey stick." The name described the shapes of the lines on the charts. They showed prices staying low and flat for a while, then rising sharply upwards back to the levels of the recent "good old days" or even higher.

Within five years, industry planners ran out of patience with the theoretical hopes as oil stayed low. More complicated gas markets also fell – slowly but even farther, from about $3 per thousand cubic feet at the onset of market deregulation in 1985 into a $1 bargain-basement cellar by the late eighties. By the start of the nineties, executives were overheard growling "they're just picking numbers out of the phone book" as they walked out on economists' speeches at trade conferences.

In 1991, lean times were accepted as reality for any foreseeable future. There was even an elaborate official declaration. This came in the form of a report, *Canadian Upstream Oil & Gas Industry Profitability*, that the Canadian Petroleum Association and Independent Petroleum Association of Canada teamed up to commission from a hard-eyed investment house: PowerWest Financial Ltd. The final nail in the coffin of the hockey-stick forecast was the Gulf War between the United Nations and Iraq over the armed attempt to annex Kuwait. In the first few months after the invasion in August of 1990, oil prices rebounded into the high $20 US range, as commodity traders speculated that the world market could not withstand a withdrawal of supplies from two OPEC countries. As soon as the UN started shooting in early 1991, traders drove the price back down $10 in a single day when it became obvious Iraq would inflict little damage in return and that plenty of oil remained available.

PowerWest took the acceptance of the new reality a step farther. On behalf of stock markets and investors that had repeatedly dealt out poundings to oil shares, the financial house marshalled volumes of grim statistics and read the riot act to corporate managers. Among the findings:

By all measures and interpretations, the oil and gas upstream industry has been unprofitable since the mid-1980s ... Returns generated are far below the industry's own historical cost of capital . . . Industry returns are inadequate relative to other sectors of the Canadian economy ... Industry returns in Canada do not compare well to international oil and gas returns . . . Overly optimistic price forecasts and the healthy cash flows generated by the industry in the early 1980s led to massive and, in hindsight, inappropriately high levels of capital spending . . . The owners of the industry, the equity holders, have suffered from inadequate returns . . . . The issue of profitability must be a central consideration by all of those who have a vested interest in the oil and gas sector.

Managers took the message to heart – to a point where the industry stood in danger of gaining a reputation as "heartless" by the mid-1990s. Mergers, asset sales, spending restraints, hiring freezes and even mass layoffs by the senior companies continued, even though oil prices stabilized and gas rebounded into the $2 range. The industry got a sharp reminder that it needs *people* (as well as rigs, pumps, pipes and computers). This time a survey, *Human Resources in the Upstream Oil & Gas Industry* (dated September 1992 but not publicly released until months later), concentrated on the consequences of single-minded dedication to improving corporate bottom lines. It was sponsored by Employment Canada, written by Peat Marwick Stevenson & Kellog and Ziff Energy Group and directed by a steering committee representing a cross-section of the industry. Among the findings:

The oil patch is no longer generally regarded as a good place to work. People who were once driven by challenge and opportunity are now driven by fear. Unless retention strategies are developed quickly, the most mobile — that is, the better qualified, motivated and younger staff — will leave the industry when the recession ends . . . If present human resource practices continue, then no amount of public relations effort will be able to reverse the image of a declining industry that will continue to discard employees who thought they had a lifetime career and are now unsuited for other jobs.

The industry acknowledged the message but made no drastic changes. It remained no harder on its rank and file than economic conditions remained on it. Both the PowerWest and Employment Canada studies described a problem deep within the industry's "fundamentals." It was expressed with a single word: "maturing." It stood for an overwhelming trend. Oil reserves were depleting across the industry's birthplace, the Western Canada Sedimentary Basin. Drilling targets were becoming smaller, harder to find and more expensive to pursue as geologists ran out of virgin formations. Between 1970 and 1990, conventional oil reserves declined by 41 percent. Persistently poor international oil prices put Arctic, offshore and oil sands development prospects beyond reach except when governments intervened with part-ownership and outright subsidies to use the industry for regional job-creation strategies.

Yet maturing has turned out to mean anything but fading away. Instead, in practice it means more than paring down assets and staff to squeeze out the last drop of profits out of dwindling reserves. While announcements of ever deeper austerity continued eight years after the fall of oil and gas prices, so did expansion on new frontiers of technology, markets and geography.

While falling short of transformations like the inventions of the automobile or pipelines, work on technological frontiers steadily brought more oil into reach. Pioneers like CS Resources Ltd. showed the way, by redefining the role of oil companies as "exploitation" rather than "exploration," then applying new techniques. By early 1991, CS earned an enthusiastic endorsement from stock analysts as a "technology play" by adapting horizontal drilling developed by one of its partners, France's state-owned Institut Français du Petrole. The technique raises production and lowers costs of wells by putting them sideways across oil reservoirs. The technique has proliferated across the united States and Canada since 1988, when diminutive CS was the second-biggest user of the technique in the world with 10 wells.

On the markets, the focus switched to natural gas. The production sector passed an historic landmark in November of 1989, when Imperial Oil Ltd. converted its Alberta cornerstone. After producing about 225 million barrels of oil at rates of up to 35 000 barrels per day since 1947, Imperial began winding down the Leduc field. The company started "blowing down" or producing a 275-billion cubic foot "cap" of natural gas that had been preserved to maintain enough pressure in the formation to keep the oil flowing. By 1994, gas accounted for about 60 percent of well completions across Western Canada.

The transportation sector fueled the switch to gas. It built new pipeline facilities that, combined with the abolition of export volume and price controls, led to sales to the US tripling between 1986 and 1993, to about 2.3 trillion cubic feet annually. TransCanada PipeLines Ltd. stood alone in expansion projects, spearheaded by capacity additions to open up new sales outlets across New York, New England and New Jersey with a brand new export route, the Iroquois Gas Transmission System. The Foothills-Northern Border route to the Middle West steadily increased delivery capacity by adding compressor power. The Alberta-to-California pipeline of Alberta Natural GAs Co., Pacific Gas Transmission Co. and Pacific Gas & Electric Co. raised its delivery capacity 75 percent and extended its reach to Los Angeles from its original destination of the San Francisco region.

Matching capacity expansions were built by the provincial gas-gathering systems of NOVA Corp. in Alberta, Westcoast Energy Inc. in British Columbia and TransGas Ltd. in Saskatchewan. To keep pace with an increasingly fast-moving market, marketing companies proliferated. Along with these, "gas-on-demand" trading services flourished, including warehouse-like storage facilities and computerized networks for rapid commodity and transportation transactions.

The gas-export surge lit a fire under drilling. By 1994, the Canadian Association of Oilwell Drilling Contractors predicted an imminent recovery to the pace of 12 000 wells a year, last seen before the oil-price crash. The heirs to Arctic prophet John Campbell (Cam) Sproule, in Sproule Associates Ltd., pointed out why in a survey conducted to support an application for further expansion submitted by TransCanada to the National Energy Board. The industry had little choice but to come back to life in the field just to make sure it kept enough supplies available to fulfil its contracts. The Sproule organization observed that the production sector can meet all the commitments, but only by satisfying a tall order. To satisfy forecast requirements of 5.8 trillion cubic feet a year across Canada and the U.S., four-fifths of production will have to come from new reserves additions by 2 015.

The call on gas reserves has become strong enough for the Sproule firm – which is renowned as conservatively realistic in its consulting role – to predict a start on development of a new supply source. Even without government subsidies akin to incentives in the U.S., Sproule Associates predicted Canadians will make a start on tapping the gas associated with coal seams. A 12-members consortium, including the Geological Survey of Canada and a *who's who* of producers, spent about $35 million of its own money experimenting with 80 coalbed methane wells by 1994. The NEB was told the group, called Canadian Coalbed Methane Forum, is showing it will have the technology to devise "an economic supplement to conventional supplies of natural gas within the next 10 to 15 years."

A growing cross-section of the industry goes beyond technology and gas to carve out new geographical frontiers abroad. Foreign work, long a fixture of an industry led by international companies, became an important part of the Canadian oil scene in the 1980s. Scores of consulting firms, service companies and roving experts worked on contracts awarded by Petro-Canada International Assistance Corp., financed by the federal government's foreign-aid budget and managed by Petro-Canada.

As the lean 1990s dawned, the aid agency was abolished as a deficit-cutting measure. By this time, commercial opportunities were developing in the biggest oil-producing nation on earth. The demise of Communism and the dismantling of the USSR lured a host of Canadian companies out to Russia and Kazakhstan for high-profile partnerships.

Siberia became the cornerstone for growth for Canadian Fracmaster Ltd., which evolved from an Alberta oil field service firm fallen on hard times into a major oil producer with its name on one of Calgary's bigger office towers. An industry cross-section, ranging from traditional giants like Gulf Canada Resources Ltd. to newcomers such as Hurricane Hydrocarbons Inc. and Cana-Kaz Global Oils Inc., lined up for joint ventures using a technique pioneered by Fracmaster. The Russians, acknowledging the ruble had little or no value, agreed to pay their foreign partners in oil and let them export it in exchange for "hard" currency such as U.S dollars.

The foreign expansion drive rapidly spread beyond oil and Russia as soon as the gas industry saw limits on the horizon to further growth in North America. The pipeline companies led the movement, observing that neither the U.S. market nor Canadian production can handle much more transportation capacity until economic conditions improve enough to warrant Arctic and offshore development. TransCanada recruited former federal energy minister Jake Epp as a vice-president for international affairs and rapidly landed agreements to work on projects proposed in Vietnam, Qatar, Pakistan and Mexico. NOVA's Novacorp International Consulting Inc., a veteran of consulting work around the globe from Turkey to Malaysia, embarked on more aggressive expansion by buying into pipelines in Argentina and Australia, with vows to look for more opportunities. Westcoast declared intentions to get into the new game, as a partner in a gas and electricity consortium that includes B.C. government agencies.

On a larger scale, the corporations' leaders reached deep to find the same kind of drive that laid-off employees had to plumb in order to survive the hard time in Western Canada. Calgarian Ron Bullen personifies the impulse. After founding Fracmaster, then leading it into Russia as a survival tactic only to be edged out by a takeover after

Siberia started paying off, he refused to retire. Recruiting son Brent as a partner, Bullen started over again as Valens Developments Ltd. Within three years, it was a name to be reckoned with in Siberia. Valens went beyond production into building houses and a 108-room, Canadian-style motel known as "the palace" in the regional industrial capital of Khanty-Mansiysk. The elder Bullen boils the petroleum industry's gritty formula for business and personal survival into a single word: "re-engage."

TransCanada chairman Gerald Maier, like Bullen a veteran of decades of petroleum's booms and busts, has made it plain senior leaders still harbor the sense of responsibility and ambition for the community that was easier to show in fatter times. In explaining why his company decided to risk trying foreign projects, Maier laid out a mandate for a bigger future on a global scale for the industry..

"If we are going to grow and keep strong and efficient, we have to use our strengths – our capital, technology, intelligence, knowledge, people and experience. Why shouldn't we keep our people fully occupied and challenged? Many countries are emerging with economic freedom. To me, it would be negligent on our part if we didn't try to avail ourselves of the opportunities that are showing up."

Pilgrimage: Dressed for August, 1985 at 76° 14′ N, energy minister Pat Carney (left) hears from Panarctic president Charles Hetherington (far right) about results of the northern partnership between government and industry. The visit inaugurated tanker shipments from Bent Horn, on Cameron Island. (Panarctic Oils Ltd.)

Further Reading

Barr, John J. "The Impact of Oil on Alberta: Retrospect and Prospect," in A.W. Rasporich, ed., *The Making of the Modern West: Western Canada Since 1945*. Calgary: University of Calgary Press, 1984.

Barry, P.S. *The Canol Project: Adventure of the U.S. War Department in Canada's Northwest*, Edmonton: P.S. Barry, 1985.

Berger, Thomas R. *Northern Frontier, Northern Homeland: The Report of the Mackenzie Valley Pipeline Inquiry*. Ottawa: Minister of Supply and Services, 1977, Vol. 1.

Bragha, François. *Bob Blair's Pipeline: The Business and Politics of Northern Energy Development Projects: New Edition with the Story Behind the Alberta Prebuild*. Toronto: James Lorimer & Company, Publishers, 1979.

Breen, David. *Alberta's Petroleum Industry and the Conservation Board*. Calgary: University of Alberta Press, 1993.

Broadfoot, Barry and Mark Nichols. *Memories: The Story of Imperial Oil's First Century as Told by its Employees and Annuitants*. Toronto: Imperial Oil Limited, 1980.

Comfort, Darlene J. *The Abasand Fiasco*. Ft. McMurray: Darlene J. Comfort, 1980.

de Mille, George. *Oil in Canada West: The Early Years*. Calgary: George de Mille, 1969.

Hanson, Eric J. *Dynamic Decade*. Toronto: McClelland and Stewart Limited, 1958.

Hilbourn, James D., consulting ed. *Dusters and Gushers: The Canadian Oil and Gas Industry*. Toronto: Pitt Publishing Company Limited, 1968.

Kerr, Aubrey. *Atlantic 1948 No. 3*. Calgary: S.A. Kerr, 1986.

Kilbourn, William. *Pipeline: Transcanada and the Great Debate: A history of business and politics*. Toronto, Vancouver: Clarke, Irwin and Company Limited, 1970.

Lewington, Peter. *No Right-of-Way: How Democracy Came to the Oil Patch*. Markham, Ontario: Fitzhenry & Whiteside, 1991.

Nickle's Daily Oil Bulletin: 50th Anniversary Edition. Calgary, Nickle's Daily Oil Bulletin: 1988.

Ross, Victor. *Petroleum in Canada*. Toronto: Southam Press Limited, 1917.

Stenson, Fred. *Waste to Wealth: A History of Gas Processing in Canada*. Calgary: Canadian Gas Processors Association/Canadian Gas Processors Supplier's Association, 1985.

Index